Sewage Sludge: Land Utilization and the Environment

Sewage Sludge: Land Utilization and the Environment

Editors
C. E. Clapp, W. E. Larson, and R. H. Dowdy

Managing Editor
David M. Kral

Associate Editor
Marian K. Viney

11–13 August 1993
Sheraton Airport Inn
Bloomington, MN

Sponsored by the
Department of Soil Science and USDA-Agricultural Research Service
University of Minnesota
St. Paul, MN

SSSA Miscellaneous Publication

Published by:
American Society of Agronomy, Inc.
Crop Science Society of America, Inc.
Soil Science Society of America, Inc.
Madison, WI, USA

1994

Cover photo provided by the Department of Soil Science at the University of Minnesota and the Metropolitan Waste Control Commission.

Cover Design: Patricia Scullion

American Society of Agronomy, Inc.
Crop Science Society of America, Inc.
Soil Science Society of America, Inc.
677 South Segoe Road, Madison, WI 53711, USA

Library of Congress Cataloging-in-Publication Data

Sewage sludge : land utilization and the environment : 11–13 August 1993, Sheraton Airport Inn, Bloomington, MN / editors, C.E. Clapp, W.E. Larson, and R.H. Dowdy : managing editor, David M. Kral : associate editor, M.K. Vinney.
p. cm. —— (SSSA miscellaneous publication)
"Sponsored by the Department of Soil Science and USDA-Agricultural Research Service, University of Minnesota, St. Paul, MN."
Includes bibliographical references.
ISBN 0-89118-813-4
1. Sewage slugde as fertilizer—Congresses. 2. Sewage sludge as fertilizer—Environmental aspects—Congresses. I. Clapp, C. E. II. Larson, William E., 1921- . III. Dowdy, R. H. IV. University of Minnesota. Dept. of Soil Science. V. United States, Agricultural Research Service. VI. Series.
S657.S48 1994
631.8′69—dc20 94-29183
CIP

Printed in the United States of America

CONTENTS

FOREWORD

The development of a sustainable society requires alternative options to landfilling and incineration as a means of waste disposal. Many states have developed goals to divert up to 40% or more of domestic wastes from landfills. Sewage sludge is a waste resource of rather consistent quality. It is a nutrient resource and organic residue that can be safely recycled provided management practices are in place to protect human health and the environment. These practices must control application rates, cumulative amounts of sludge applied and appropriate dosing periods between application and grazing or crop harvest. Soil and crop scientists, including many of those who authored papers for this publication, have been instrumental in developing the knowledge base and providing the expertise for practices and regulations governing sewage sludge utilization on the land. The federal regulations on sewage sludge applications are perhaps the most technically based set of regulations ever developed to satisfy an environmental concern. Much progress has been made in utilizing sludge as a by-product resource, while protecting human health and the environment. More research will probably be needed, however, to meet society mandates for the next decade. The authors and editors are to be commended for bringing this text to fruition in a timely manner. It will be well received by professionals engaged in managing waste products in near surface earth systems at a period in our history when sustainability of the biosphere is a national and international priority.

L. P. WILDING, *president*
Soil Science Society of America

PREFACE

Utilization of sewage wastes on land for enhancement of crop production is an age-old practice. Treatises on the subject were written as far back as the time of the Roman Empire. Prior to modern times, organic wastes were usually applied directly to the land without processing, although some wastes may have been composted. In the current century with more concern for health hazards from raw wastes, direct application was largely discontinued. Instead wastewater treatment plants were built for both primary and secondary treatment. The water from the plants was usually discharged into rivers and the sludge was discharged into large water bodies, dumped into landfills or incinerated. Small municipalities with limited facilities often continued to apply the sludge to land, although often it was disposed of with little regard for its agricultural value.

In the last few decades concern for environmental contamination prompted a rethinking of the concept of utilization of sludge on land. Agricultural utilization of sewage sludge has the potential benefit of utilization of the nutrients and organic matter for crop production and soil management. Added benefits are saving in energy consumption, prevention of air pollution in the case of incineration and reduction in water contamination in the case of ocean or lake dumping. But until recently, inadequate information existed for safe, efficient use of sludge on land for crop production.

Scientists and engineers responded to the need for research information on efficient and environmentally safe utilization of sewage sludge on land. A National Land Utilization of Sludge Conference was held at the University of Illinois in 1973. A second national conference was held in Colorado in 1983. These conferences were followed by a third conference in Minnesota in 1993. The Proceedings of the 1993 conference is the subject of this book.

Marked technical progress concerning land utilization of sludge has been made in the past two decades. Building on previous work, the papers in this Proceedings bring the past and current research information together into one volume. Included are 22 invited papers and 10 volunteer poster papers. Also included is a section on the most pressing research needs.

On behalf of all attendants, we wish to express our gratitude to all sponsoring organizations for their support and to ASA-CSSA-SSSA for publishing this book. We also wish to express our appreciation to all speakers and poster participants for their excellent presentations and manuscripts and to all contributors who made the conference a success. Special thanks go to the session chairs and manuscript reviewers for their many helpful suggestions.

C. E. CLAPP, co-editor
USDA-ARS, University of Minnesota, St. Paul, MN

W. E. LARSON, co-editor
University of Minnesota, St. Paul, MN

R. H. DOWDY, co-editor
USDA-ARS, University of Minnesota, St. Paul, MN

ACKNOWLEDGMENTS

Organizing Committee

H. H. Cheng, Chair
C. E. Clapp
R. H. Dowdy
T. R. Halbach
W. E. Larson
S. A. Stark

Editorial and Review Committee

J. R. Brown
H. H. Cheng
C. E. Clapp, Chair
R. H. Dowdy
S. J. Henning
W. E. Larson
D. R. Linden
T. J. Logan
A. E. Peterson
S. A. Stark
D. G. Westfall

Co-Sponsors:

Association of Metropolitan Sewerage Agencies

Metropolitan Waste Control Commission, St. Paul

N-Viro International, Inc.

N-Viro Minnesota, Inc.

Soil Science Society of America

University of Minnesota, Department of Soil Science

U. S. Army Corps of Engineers - CRREL

U. S. Department of Agriculture - Agricultural Research Service

Arrangements for the conference, including mailings, registration, program preparation, and facilities and field tour organization were coordinated by the University of Minnesota Extension Service's Educational Development System, under the capable direction of Nancy Harvey and Eugene Anderson. Many of the details associated with the conference sessions and field tour were handled by students and support staff of the Department of Soil Science. Preparation of manuscript drafts and final camera-ready copies of the proceedings publication were professionally carried out by Karen Mellem.

RESEARCH NEEDS

Research Needs for Land Application

A greatly expanded research program on land application of sewage sludge was started about two decades ago in the USA. At that time only a small fraction of domestic sludge was land applied and little was known about best management practices. This interest in land application was spurred by a more intensive recognition of the need to protect the environment, and, to a lesser extent, the need to conserve energy resources. The expanded interest was marked by a national land utilization of sewage sludge conference held at the University of Illinois in 1973. A second conference to assess the research progress was held in Denver in 1983. A third national conference was held in 1993 in Minnesota. The Proceedings of the 1993 conference presents the status of our current research information.

A great deal of progress has been made in research leading to safe, practical methods of land utilization of municipal sewage sludge as is so well presented in the papers of this 1993 conference. The research done by scientists and engineers from a variety of institutions and representing many disciplines is a true success story. But, still unanswered questions need further research. Each invited author was asked to include in their papers a section on research needs. These were discussed at the workshop and a committee summarized the most pressing needs in an attempt to help guide future research activities. These research needs follow:

Long-term Studies

Long-term (decades) studies are needed to determine slow, but real, changes in soil and plant composition effected by continuous applications of sludge-borne pollutants. Changes such as soil binding capacity as the soil and sludge attributes interact and the effect of aging on the susceptibility for transport need to be documented for both agricultural and natural ecosystems. The hypothesis that trace element activity in soil-sludge mixtures is sludge-controlled, once a critical level has been exceeded for a given sludge, needs field testing.

Beneficial Effect

Inadequate fundamental knowledge is available for predicting beneficial effects obtainable after sludge and sludge-compost applications. Interactions between soil organic matter decomposition level, sludge- and soil-borne plant pathogens and plant roots must be explored further. Better, more reliable methods for predicting nitrogen (N) mineralization rates for different sludges and admixtures of sludge and alternate carbon (C) and mineral sources are required for minimizing the potential for nitrate (NO_3) leaching.

Risks

The risks from introduction of toxic organics, trace metals, and pathogens into the farm ecosystem and possible introduction into the food chain, and water resources for both humans, and agricultural and wildlife animals need further elucidation. It is vitally important to provide information on uncertainty associated with risk assessment and the size of the population at risk. New methods must be developed that include, but differentiate, among sources of uncertainty and are capable of displaying the uncertainty. This includes use of statistical distributions of parameters to develop probability distributions of exposure rather than reliance on point estimates and the highly exposed individual (HEI) approach. From this knowledge, rational criteria can be established for land application of sewage sludge.

Systems Research

Systems research is needed to define viable, management options for the farm operator. Synthesis of current information and new research into practical, easily understood management systems could make sludge usage on farms more acceptable. The risks and benefits need to be clearly defined. Responsibilities and liabilities of farm operators and sewage agencies must be spelled out.

Trace Elements

The shape of the rate response curve for trace metal uptake by plants needs to be confirmed with field data. The assumption that plant response is relative (all crops can be represented by the response of a single crop) must be field tested and validated. The bio-availability of metals in plant tissues, sludges, and soil-sludge mixtures ingested by livestock and wildlife needs to be fully characterized. This includes the case in which earthworm-consuming mammals and birds ingest high amounts of soil. A better understanding of pollutant desorption is needed to support models that consider transport in non-uniform media.

Ecological Effects

Information is needed on the ecological effects of organic and inorganic constituents of sludge on wildlife, water resources, and noncultivated crops and impacts on unmanaged plant and animal communities. The impacts on adjacent plant and animal communities and the distances over which impacts may occur needs study.

Design, Use, and Cost Considerations

A need exists to develop a concise quantitative understanding of the magnitude of sludge generation, and the impact of various alternatives on the costs of sludge utilization. Various design alternatives are possible. Processes in wastewater treatment plants as well as post plant-treatment handling alternatives need careful study. A second National Sewage Sludge Survey is needed to update the concentrations of constituents in sludge and to determine what changes have occurred in sludge management practices.

Sociological–Economic: Marketing Considerations

Reusing wastes in a beneficial way is not an integral part of U.S. culture. As a result, recycling of waste into agriculture raises many concerns, both real and imagined. A rational use policy needs to provide economic compensation for property impacts, both real or imagined. The impact of the practice of sludge use in food production and its impact on international trade needs to be carefully considered. This may require establishing international sludge surveys that include background levels of various microbiological and inorganic constituents not only in sludge but also in foods.

Technical Specifications for Sewage Sludge and Similar Products

Sewage sludge, animal manure, municipal solid wastes, composts, green waste, and other organic amendments are all applied on land. Benefits of using these materials, however, are dependent upon properties of the applied products. At present, marketing and promotion of their uses on land are hampered because their chemical and physical properties vary considerably, even among the same type of material. Technical specifications need development. These specifications will serve as industry-wide standards for product performance and will be helpful to establish consumer confidence in the products.

CONTRIBUTORS

G. H. Abd El-Hay	Faculty of Agriculture, Al-Azhar University, Nasr City, Cairo, Egypt
Richard Aguilar	Rocky Mountain Forest and Range Experiment Station, USDA-Forest Service, Albuquerque, NM 87106
J. Scott Angle	Dep. of Agronomy, University of Maryland, College Park, MD 20742
D. A. J. Barry	Centre for Land and Water Stewardship, University of Guelph, Guelph, Ontario N1G 2W1, Canada
R. K. Bastian	Office of Wastewater Enforcement and Compliance, U.S. Environmental Protection Agency, Washington, DC 20460
S. L. Brown	Dep. of Agronomy, University of Maryland, College Park, MD 20742
S. Burke	Dep. of Environmental Sciences, Allegheny College, Meadville, PA 16335
Rufus L. Chaney	USDA-Agricultural Research Service, Beltsville, MD 20705
A. C. Chang	Dep. of Soil and Environmental Sciences, University of California, Riverside, CA 92521-0424
H. H. Cheng	Dep. of Soil Science, University of Minnesota, St. Paul, MN 55108-6028
C. E. Clapp	USDA-Agricultural Research Service, University of Minnesota, St. Paul, MN 55108-6028
Dale W. Cole	College of Forest Resources, University of Washington, Seattle, WA 98105
R. B. Corey	Dep. of Soil Science, University of Wisconsin, Madison, WI 53706

W. Lee Daniels	Dep. of Crop and Soil Environmental Sciences, Virginia Polytechnic Institute and State University, Blacksburg, VA 24061-0404
R. H. Dowdy	USDA-Agricultural Research Service, University of Minnesota, St. Paul, MN 55108-6028
D. R. Duncomb	Dep. of Soil Science, University of Minnesota, St. Paul, MN 55108-6028
Anne Fairbrother	Environmental Research Laboratory, U.S. Environmental Protection Agency, Corvallis, OR 97333
E. Filcheva	Poushkarov Institute of Soil Science and Agroecology, Sofia, Bulgaria
Jane B. Forste	Bio Gro Systems, Annapolis, MD 21404
Philip R. Fresquez	Los Alamos National Laboratory, Los Alamos, NM 87544
A. Garcia	Dep. of Agriculture, Prairie View A&M University, Prairie View, TX 77446
P. Brad Gates	Dep. of Soil, Crop and Atmospheric Sciences, Cornell University, Ithaca, NY 14853
D. Goorahoo	Centre for Land and Water Stewardship, University of Guelph, Guelph, Ontario N1G 2W1, Canada
M. J. Goss	Centre for Land and Water Stewardship, University of Guelph, Guelph, Ontario N1G 2W1, Canada
T. C. Granato	Research and Development Department, Metropolitan Water Reclamation District of Greater Chicago, Chicago, IL 60611
C. E. Green	USDA-Agricultural Research Service Beltsville, MD 20705

J. Gschwind — Research and Development Department, Metropolitan Water Reclamation District of Greater Chicago, Chicago, IL 60611

Kathryn C. Haering — Dep. of Crop and Soil Environmental Sciences, Virginia Polytechnic Institute and State University, Blacksburg, VA 24061-0404

T. R. Halbach — Dep. of Soil Science, University of Minnesota, St. Paul, MN 55108-6028

Charles L. Henry — College of Forest Resources, University of Washington, Seattle, WA 98105

Harry A. J. Hoitink — Dep. of Plant Pathology, Ohio Agric. Research Center, The Ohio State University, Wooster, OH 44691

Christara M. Hormann — Dep. of Soil Science, University of Minnesota, St. Paul, MN 55108-6028

Iskandar K. Iskandar — U.S. Army Cold Regions Research and Engineering Laboratory, Hanover, NH 03755

J. A. Jeffery — USDA-Soil Conservation Service, Brooklyn Center, MN 55429

William J. Jewell — Dep. of Agricultural and Biological Engineering, Cornell University, Ithaca, New York 14853

D. R. Keeney — Dep. of Agronomy, Iowa State University, Ames, IA 50011

C. M. Knapp — Technical Resources, Davis, CA 95616

W. E. Larson — Dep. of Soil Science, University of Minnesota, St. Paul, MN 55108-6028

D. R. Linden — USDA-Agricultural Research Service, University of Minnesota, St. Paul, MN 55108-6028

Samuel R. Loftin	Rocky Mountain Forest and Range Experiment Station, USDA-Forest Service, Albuquerque, NM 87106
Cecil Lue-Hing	Research and Development Department, Metropolitan Water Reclamation District of Greater Chicago, Chicago, IL 60611
A. S. Mangaroo	Dep. of Agriculture, Prairie View A&M University, Prairie View, TX 77446
G. A. O'Connor	Dep. of Soil and Water Science, University of Florida, Gainesville, FL 32611
S. L. Oberle	Dep. of Agronomy, Iowa State University, Ames, IA 50011
A. L. Page	Dep. of Soil and Environmental Sciences, University of California, Riverside, CA 92507
Antonio J. Palazzo	U.S. Army Cold Regions Research and Engineering Laboratory, Hanover, NH 03755
E. Pallant	Dep. of Environmental Sciences, Allegheny College, Meadville, PA 16335
A. E. Peterson	Dep. of Soil Science, University of Wisconsin, Madison, WI 53706
John H. Peverly	Dep. of Soil, Crop and Atmospheric Sciences, Cornell University, Ithaca, NY 14853
Gary M. Pierzynski	Dep. of Agronomy, Kansas State University, Manhattan, KS 66506
R. I. Pietz	Research and Development Department, Metropolitan Water Reclamation District of Greater Chicago, Chicago, IL 60611
R. C. Polta	Metropolitan Waste Control Commission, St. Paul, MN 55106

James A. Ryan — Risk Reduction Engineering Laboratory, U.S. Environmental Protection Agency, Cincinnati, OH 45202

P. L. Schlecht — Milwaukee Metropolitan Sewerage District, Oak Creek, WI 53154

Kelly E. Smith — Dep. of Soil Science, University of Minnesota, St. Paul, MN 55108-6028

P. S. Smith — Centre for Land and Water Stewardship, University of Guelph, Guelph, Ontario N1G 2W1 Canada

P. E. Speth — Dep. of Soil Science, University of Wisconsin, Madison, WI 53706

John M. Walker — Office of Wastewater Enforcement and Compliance, U.S. Environmental Protection Agency, Washington, DC 20460

T. H. Wright — Dep. of Soil Science, University of Wisconsin, Madison, WI 53706

D. R. Zenz — Research and Development Department, Metropolitan Water Reclamation District of Greater Chicago, Chicago, IL 60611

Conversion Factors for SI and non-SI units

Conversion Factors for SI and non-SI Units

To convert Column 1 into Column 2, multiply by	Column 1 SI Unit	Column 2 non-SI Unit	To convert Column 2 into Column 1, multiply by
		Length	
0.621	kilometer, km (10^3 m)	mile, mi	1.609
1.094	meter, m	yard, yd	0.914
3.28	meter, m	foot, ft	0.304
1.0	micrometer, μm (10^{-6} m)	micron, μ	1.0
3.94×10^{-2}	millimeter, mm (10^{-3} m)	inch, in	25.4
10	nanometer, nm (10^{-9} m)	Angstrom, Å	0.1
		Area	
2.47	hectare, ha	acre	0.405
247	square kilometer, km^2 (10^3 m)^2	acre	4.05×10^{-3}
0.386	square kilometer, km^2 (10^3 m)^2	square mile, mi^2	2.590
2.47×10^{-4}	square meter, m^2	acre	4.05×10^3
10.76	square meter, m^2	square foot, ft^2	9.29×10^{-2}
1.55×10^{-3}	square millimeter, mm^2 (10^{-3} m)^2	square inch, in^2	645
		Volume	
9.73×10^{-3}	cubic meter, m^3	acre-inch	102.8
35.3	cubic meter, m^3	cubic foot, ft^3	2.83×10^{-2}
6.10×10^4	cubic meter, m^3	cubic inch, in^3	1.64×10^{-5}
2.84×10^{-2}	liter, L (10^{-3} m^3)	bushel, bu	35.24
1.057	liter, L (10^{-3} m^3)	quart (liquid), qt	0.946
3.53×10^{-2}	liter, L (10^{-3} m^3)	cubic foot, ft^3	28.3
0.265	liter, L (10^{-3} m^3)	gallon	3.78
33.78	liter, L (10^{-3} m^3)	ounce (fluid), oz	2.96×10^{-2}
2.11	liter, L (10^{-3} m^3)	pint (fluid), pt	0.473

	Mass		
2.20×10^{-3}	gram, g (10^{-3} kg)	pound, lb	454
3.52×10^{-2}	gram, g (10^{-3} kg)	ounce (avdp), oz	28.4
2.205	kilogram, kg	pound, lb	0.454
0.01	kilogram, kg	quintal (metric), q	100
1.10×10^{-3}	kilogram, kg	ton (2000 lb), ton	907
1.102	megagram, Mg (tonne)	ton (U.S.), ton	0.907
1.102	tonne, t	ton (U.S.), ton	0.907
	Yield and Rate		
0.893	kilogram per hectare, kg ha^{-1}	pound per acre, lb $acre^{-1}$	1.12
7.77×10^{-2}	kilogram per cubic meter, kg m^{-3}	pound per bushel, lb bu^{-1}	12.87
1.49×10^{-2}	kilogram per hectare, kg ha^{-1}	bushel per acre, 60 lb	67.19
1.59×10^{-2}	kilogram per hectare, kg ha^{-1}	bushel per acre, 56 lb	62.71
1.86×10^{-2}	kilogram per hectare, kg ha^{-1}	bushel per acre, 48 lb	53.75
0.107	liter per hectare, L ha^{-1}	gallon per acre	9.35
893	tonnes per hectare, t ha^{-1}	pound per acre, lb $acre^{-1}$	1.12×10^{-3}
893	megagram per hectare, Mg ha^{-1}	pound per acre, lb $acre^{-1}$	1.12×10^{-3}
0.446	megagram per hectare, Mg ha^{-1}	ton (2000 lb) per acre, ton $acre^{-1}$	2.24
2.24	meter per second, m s^{-1}	mile per hour	0.447
	Specific Surface		
10	square meter per kilogram, m^2 kg^{-1}	square centimeter per gram, cm^2 g^{-1}	0.1
1000	square meter per kilogram, m^2 kg^{-1}	square millimeter per gram, mm^2 g^{-1}	0.001
	Pressure		
9.90	megapascal, MPa (10^6 Pa)	atmosphere	0.101
10	megapascal, MPa (10^6 Pa)	bar	0.1
1.00	megagram per cubic meter, Mg m^{-3}	gram per cubic centimeter, g cm^{-3}	1.00
2.09×10^{-2}	pascal, Pa	pound per square foot, lb ft^{-2}	47.9
1.45×10^{-4}	pascal, Pa	pound per square inch, lb in^{-2}	6.90×10^3

(continued on next page)

Conversion Factors for SI and non-SI Units

To convert Column 1 into Column 2, multiply by	Column 1 SI Unit	Column 2 non-SI Unit	To convert Column 2 into Column 1, multiply by
		Temperature	
1.00 (K − 273)	Kelvin, K	Celsius, °C	1.00 (°C + 273)
(9/5 °C) + 32	Celsius, °C	Fahrenheit, °F	5/9 (°F − 32)
		Energy, Work, Quantity of Heat	
9.52×10^{-4}	joule, J	British thermal unit, Btu	1.05×10^{3}
0.239	joule, J	calorie, cal	4.19
10^{7}	joule, J	erg	10^{-7}
0.735	joule, J	foot-pound	1.36
2.387×10^{-5}	joule per square meter, $J\ m^{-2}$	calorie per square centimeter (langley)	4.19×10^{4}
10^{5}	newton, N	dyne	10^{-5}
1.43×10^{-3}	watt per square meter, $W\ m^{-2}$	calorie per square centimeter minute (irradiance), $cal\ cm^{-2}\ min^{-1}$	698
		Transpiration and Photosynthesis	
3.60×10^{-2}	milligram per square meter second, $mg\ m^{-2}\ s^{-1}$	gram per square decimeter hour, $g\ dm^{-2}\ h^{-1}$	27.8
5.56×10^{-3}	milligram (H_2O) per square meter second, $mg\ m^{-2}\ s^{-1}$	micromole (H_2O) per square centimeter second, $\mu mol\ cm^{-2}\ s^{-1}$	180
10^{-4}	milligram per square meter second, $mg\ m^{-2}\ s^{-1}$	milligram per square centimeter second, $mg\ cm^{-2}\ s^{-1}$	10^{4}
35.97	milligram per square meter second, $mg\ m^{-2}\ s^{-1}$	milligram per square decimeter hour, $mg\ dm^{-2}\ h^{-1}$	2.78×10^{-2}
		Plane Angle	
57.3	radian, rad	degrees (angle), °	1.75×10^{-2}

	Electrical Conductivity, Electricity, and Magnetism		
10	siemen per meter, S m^{-1}	millimho per centimeter, mmho cm^{-1}	0.1
10^4	tesla, T	gauss, G	10^{-4}
	Water Measurement		
9.73×10^{-3}	cubic meter, m^3	acre-inches, acre-in	102.8
9.81×10^{-3}	cubic meter per hour, m^3 h^{-1}	cubic feet per second, ft^3 s^{-1}	101.9
4.40	cubic meter per hour, m^3 h^{-1}	U.S. gallons per minute, gal min^{-1}	0.227
8.11	hectare-meters, ha-m	acre-feet, acre-ft	0.123
97.28	hectare-meters, ha-m	acre-inches, acre-in	1.03×10^{-2}
8.1×10^{-2}	hectare-centimeters, ha-cm	acre-feet, acre-ft	12.33
	Concentrations		
1	centimole per kilogram, cmol kg^{-1}	milliequivalents per 100 grams, meq 100 g^{-1}	1
0.1	gram per kilogram, g kg^{-1}	percent, %	10
1	milligram per kilogram, mg kg^{-1}	parts per million, ppm	1
	Radioactivity		
2.7×10^{-11}	becquerel, Bq	curie, Ci	3.7×10^{10}
2.7×10^{-2}	becquerel per kilogram, Bq kg^{-1}	picocurie per gram, pCi g^{-1}	37
100	gray, Gy (absorbed dose)	rad, rd	0.01
100	sievert, Sv (equivalent dose)	rem (roentgen equivalent man)	0.01
	Plant Nutrient Conversion		
	Elemental	***Oxide***	
2.29	P	P_2O_5	0.437
1.20	K	K_2O	0.830
1.39	Ca	CaO	0.715
1.66	Mg	MgO	0.602

SECTION I

OVERVIEW OF THE PAST 25 YEARS

1 Overview of the Past 25 Years: Technical Perspective

A. L. Page
A. C. Chang

Department of Soil and Environmental Sciences
University of California
Riverside, California

Archeology evidence has traced the concept of domestic wastewater treatment and disposal back to antiquity. The practice of community-wide systematic collection, treatment, and disposal of wastewater, however, did not evolve until the late 19th Century. When the municipal wastewaters were directly discharged, sewage sludge did not exist. As the wastewater treatment technology advanced and the potential pollutants were effectively removed from the wastewater, the treated effluents become a lesser threat to the environment and human health. Municipal sewages are residues of wastewater treatment and the chemical constituents in the sludge represent an agglomeration of pollutants originally present in the wastewater. Sewage sludge requires ultimate disposal.

Since the beginning, land application has been a popular option for disposing of sewage sludge. Early promoters of this practice advocated the use of soil as a *treatment* medium and wastes as a source of plant nutrients. Even the municipal sludges produced a long time ago were not free of potentially harmful elements. Fifty years ago, Rudolfs and Gehm (1942) documented the agronomic attributes of municipal sludges and reported that sludges in the USA contained 160 to 400 mg kg^{-1} copper (Cu), 80 to 320 mg kg^{-1} zinc (Zn), 930 to 1860 mg kg^{-1} lead (Pb), and up to 1400 mg kg^{-1} chromium (Cr).

The nation-wide environmental movement in the early 1970s heightened our concerns on the need of proper disposal of sewage sludges. While the basic concepts and fundamental processes governing physical, chemical, and biological reactions in the soil were reasonably well understood at the time, information on the fate and effects of sewage sludge-borne contaminants in soils was largely unknown. Technical data necessary to characterize pollution potential of sewage sludge, to assess adverse impacts due to the transfer of potentially harmful chemical constituents from sludge to surface and ground water bodies, and to consumers through the food chain were either inadequate or totally lacking. From 1975 to 1993, we witnessed an intense and concerted effort of scientific

Sewage Sludge: Land Utilization and the Environment. SSSA Miscellaneous Publication.

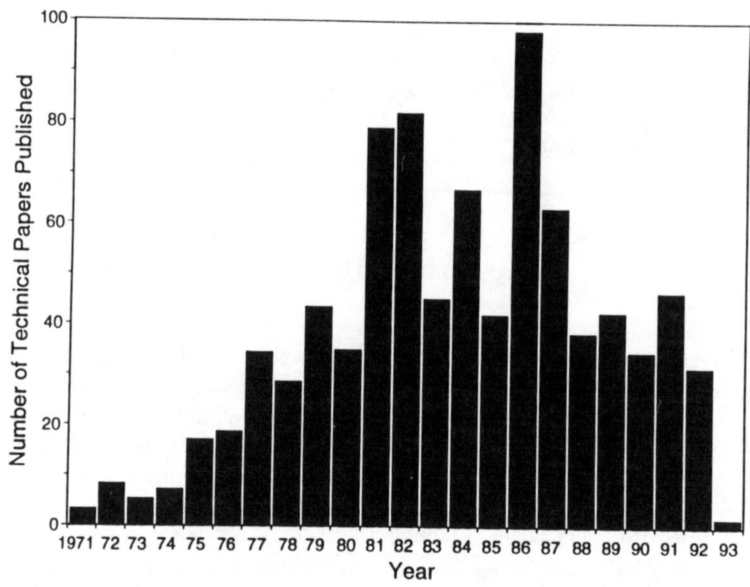

Fig. 1-1. Number of sludge land application related technical publications (1970 to March, 1993).

research world-wide to better understand reactions of sludge constituents in soils. In the USA, land application research was sponsored largely by federal agencies (USDA-CSRS, USDA-ARS, and USEPA) and publicly owned treatment works (POTWs). A casual search of the computer database AGRICOLA revealed 876 references from 1970 to March, 1993; essentially all were technical reports evaluating impacts of sludge constituents on soils, plants, and ground water quality (Fig. 1-1).

The research agenda on land application of municipal sludge was set at a 1973 workshop sponsored by a subcommittee of U.S. Environmental Protection Agency, U.S. Department of Agriculture, and the National Association of Land Grant Colleges (National Association of State Universities and Land Grant Colleges, 1974). This workshop identified the research needs for environmentally safe use of wastewater effluents and sewage sludges on land with respect to economic, engineering, and public health. Workshop participants were in agreement that sludges could be useful soil conditioners and sources of plant nutrients [nitrogen (N), phosphorus (P), and micronutrients], but information was needed to properly assess potential long-term hazards associated with its application to land. The potentials for nitrate (NO_3) leaching, food chain transfer of toxic chemicals, and phytotoxicity of trace elements were the primary concerns.

Elements most important in terms of the human food chain were arsenic (As), cadmium (Cd), mercury (Hg), Pb, selenium (Se), and Zn. Those most important in terms of phytotoxicity were boron (B), Cd, Cu, nickel (Ni), and Zn. The work group concluded that, for land application of sludge to move forward, there was a need for laboratory and greenhouse evaluation of the suitability of sludge, crops, and soils to generate the baseline data.

To prevent NO_3 leaching, it would be essential that the input of N resulting from sludge application be balanced by its output from crop uptake and loss by volatilization and denitrification. During the past 20 yr, we greatly improved our understanding of the physical, chemical and biological processes controlling N transformation in relation to soil and sludge properties. This knowledge has provided us with the tools to plan, design, and operate sludge land application systems such that N leaching is minimized.

Numerous laboratory and field studies have demonstrated that most trace elements (including common industrial metal elements) are relatively immobile in soils and once added will remain in the layer of incorporation. Therefore, problems associated with trace elements entering ground water beneath land where sludge has been applied is unlikely (Yingming & Corey, 1993). Very sandy soils underlain by shallow aquifers perhaps are an exception.

In an effort to simulate effects of trace metals in sludge-amended soils, the early investigations focused on growing potted plants in the greenhouse with metals added to soils as inorganic salts and inorganic metal salt-spiked sludges. Subsequent field studies used production-quality sludges and showed that the results of greenhouse studies grossly overestimated bioavailability and phytotoxicity of sludge-borne metal elements. Results of field studies demonstrated that crops grown on sludge-amended soils benefitted from the plant nutrients in sewage sludge, but would not accumulate As, Cr, Cu, Pb, Hg, Ni, or Zn in amounts sufficient to be harmful to consumers. Depending upon the soil conditions, however, there was a potential for Cd, molybdenum (Mo), and Se to be taken up by crops in amounts harmful to humans (Cd) and animals (Cd, Mo, and Se) who consumed the affected harvest (Logan & Chaney, 1983). Except possibly on very acid soils ($pH < 5.5$), there are no documented cases of Cu, Ni, and Zn toxicity to plants grown on sludge-amended soils.

In 1993, the questions raised by scientists during the 1973 sludge workshop have all been answered. While there are loose ends to be settled, there is adequate technical information for professionals in waste management to plan and design sewage sludge land application systems. In the USA, two decades of research effort in land application by many scientists across the nation has culminated in the promulgation of *Standards for the Disposal and Utilization of Sewage Sludge* (Code of Federal Regulations, Title 40, Parts 257 and 503). This regulation defines the pollutant loading boundary of land application (U.S. Environmental Protection Agency, 1993). The attention should now be focused on developing strategies for POTWs to implement the land application practice.

REFERENCES

Logan, T.J., and R.L. Chaney. 1983. Utilization of municipal wastewater and sludge on land: Metals. p. 235–326. *In* A.L. Page et al. (ed.) Utilization of municipal wastewater and sludge on land. Denver, CO. 23-25 Feb. 1983. Univ. of California, Riverside.

National Association of State Universities and Land Grant Colleges. 1974. Proceedings of the Joint Conference on Recycling Municipal Sludges and Effluents on Land. Champaign, IL. 9-13 July 1973. NASULGU, Washington, DC.

Rudolfs, W., and W.H. Gehm. 1942. Chemical composition of sewage sludges, with particular reference to their phosphoric acid contents. New Jersey Agric. Exp. Stn. Bull. 699. New Brunswick.

U.S. Environmental Protection Agency. 1993. Standards for the use or disposal of sewage sludge. Federal Register 58(32):9248–9415. U.S. Gov. Print. Office, Washington, DC.

Yingming, L., and R.B. Corey. 1993. Redistribution of sludge-borne cadmium, copper, and zinc in a cultivated plot. J. Environ. Qual. 2:1–8.

2 Overview of the Past 25 Years: Operator's Perspective

Cecil Lue-Hing
R. I. Pietz
T. C. Granato
J. Gschwind
D. R. Zenz

Research and Development Department
Metropolitan Water Reclamation District of Greater Chicago
Chicago, Illinois

The quantities of sewage sludge applied to land in the USA has steadily increased during the last two decades. Table 2-1 shows that in 1972, 20% of the total sludge produced in the USA went to land application and 40% went to landfills. In 1989, 33.3% of the total sludge produced went to land application as compared with 33.9% for landfills.

The quantities of sewage sludge used under various practices changed from 1972 to 1989 as shown in Table 2-2. In 1972, 0.86 million dry metric tons of sewage sludge were applied to land (Smith, 1977), and by 1989, the amount had increased to 2.34 million dry metric tons (U.S. Environmental Protection Agency, 1993). The amount of sludge landfilled in 1972 as compared with 1989 was almost the same, and the amount of sludge incinerated in 1989 was less than that incinerated in 1972.

The U.S. Environmental Protection Agency (USEPA) conducted a National Sewage Sludge Survey in 1988, and based on the results from that survey, the USEPA estimated the mass of sewage sludge disposed annually. Table 2-3 shows a summary from the survey and lists the mass of sludge used or disposed according to publicly owned treatment works (POTWs) size and the major use or disposal practices.

HURDLES TO LAND APPLICATION OF SLUDGE

During the past 25 yr, the application of sewage sludge to land had several major hurdles to overcome. Public opposition to land application by POTWs was due to concerns about odors, pathogens, heavy metals, and contamination of surface and ground waters. Also due to a lack of good research data, it was very difficult for municipalities, regulatory agencies and other technical experts to assess the risk to public health and the environment

Sewage Sludge: Land Utilization and the Environment. SSSA Miscellaneous Publication.

Table 2-1. Percentages of municipal sewage sludge used or disposed of in the USA.

Use or disposal practice	1972†	1989‡
	%	
Landfill	40	33.9
Land application	20	33.3
Surface disposal	ND§	10.3
Incineration	25	16.1
Ocean disposal	15	6.3

†Smith, 1977.
‡U.S. Environmental Protection Agency, 1993.
§No data on surface disposal were collected.

Table 2-2. Quantities of municipal sewage sludge used or disposed of in the USA.

Use or disposal practice	1972†	1989‡
	Million dry metric tons yr^{-1}	
Landfill	1.72	1.81
Land application	0.86	1.79
Surface disposal	ND§	0.55
Incineration	1.07	0.86
Ocean disposal	0.64	0.34
Total	4.29	5.35

†Smith, 1977.
‡U.S. Environmental Protection Agency, 1993.
§No data on surface disposal were collected.

Table 2-3. Estimated mass of sewage sludge managed annually in the USA by size of publicly owned treatment works (POTW) and use or disposal practice.

Use or disposal practice	Mass of sewage sludge used or disposed by POTW size[†]			
	>100 mgd[‡]	10 - 100 mgd	1 - 10 mgd	<1 mgd
	Thousands of dry metric tons			
Landfill	518.4	673.6	495.4	110.4
Land application	387.7	664.7	538.1	178.1
Surface disposal	79.5	264.6	122.1	87.2
Incineration	382.8	346.3	124.7	10.5
Ocean disposal	166.0	157.8	8.0	3.4

[†]U.S. Environmental Protection Agency, 1993.
[‡]million gallons per day.

from the land application of sewage sludge. Thus, the regulators adopted a conservative posture to ensure environmental protection.

Twenty-five years ago, landfilling and incineration were very convenient and inexpensive means of managing sludge as shown by the data in Tables 2-1 and 2-2. The lack of public acceptance for land application made land acquisition difficult and costly, and the conservative posture adopted by regulators produced restrictive criteria that increased the cost for operators. In other instances, the lack of regulations made it difficult for POTWs to establish creditable land application programs for their sewage sludge.

There were several other hurdles encountered by municipal agencies in trying to establish land application programs. A major problem was the misinformation about the behavior of sludge-borne metals in soil and their uptake by plants. There was also misinformation about the viability and persistence of sludge-borne pathogens and the effect of sludge-borne metals in soils and crops on animal and human health. Other problems included the availability of suitable land, transportation costs, and the legal and political problems associated with overland transportation of sewage sludge across jurisdictional boundaries. Because of these constraints, land application was not the option of choice for many municipalities.

POSITIVE IMPETUS TO LAND APPLICATION

Despite the major hurdles to the use of land application as a viable sludge management practice, significant progress has been made in developing successful programs. A large part of the progress has been the commitment by the USEPA to study land application during the development of the 40 CFR Part 503 sewage sludge regulations (U.S. Environmental Protection Agency, 1993). The USEPA has supported research during the past two decades to foster the development of comprehensive sludge management regulations. As a result, the USEPA can now boast a risk-based regulation that sets limits based on risk to human health and the environment.

The development of the Part 503 regulations was greatly aided by the availability of research data upon which to base the regulations. In 1973, a conference was held in Champaign, IL, to establish research priorities for land application (National Association of State Universities and Land Grant Colleges, 1974). In 1976, the Council for Agricultural Science and Technology (CAST) at the request of USEPA reviewed guidelines to limit sludge application to land and evaluated metals in sludge and their potential hazard to plants, animals, and humans (Council for Agricultural Science and Technology, 1976). In 1980, another CAST report was published that evaluated the effects of sludge-applied cadmium and zinc in soils and crops (Council for Agricultural Science and Technology, 1980). A workshop was held in Denver, CO, in 1983 to review the impact of metals in sludge on soils, crops, and animals (Logan & Chaney, 1983). In 1985, a workshop was held in Las Vegas, NV, to evaluate the effects of sewage sludge quality and soil properties on plant uptake of sludge-applied trace constituents (Page et al., 1987).

A major impact on the development of the Part 503 sludge regulations was a peer review of the draft regulations published in 1989. At the request of the USEPA, the Cooperative State Research Service Regional Technical Committee W-170, with assistance of experts from the USEPA, academia, environmental groups, and units of state and local government agencies organized a Peer Review Committee that conducted a review of the draft sludge regulations (U.S. Department of Agriculture, 1989). The Peer Review Committee was divided into work groups that evaluated monofills, surface disposal, nonagricultural land application, agricultural land application, distribution and marketing, and risk assessment.

The Part 503 sludge regulations, developed by the USEPA are risk-based and consider the effect of a pollutant on a highly exposed individual (human, plant, or animal), and was supported by an aggregate risk assessment on populations at higher risk. For land application the USEPA considered 14 pathways, imposed N control for ground water protection, imposed surface water protection features, implemented pathogen reduction and vector attraction control requirements, and established metal control requirements with limits for 10 metals as shown in Table 2-4. Land application, under the Part 503 regulations, governs use on farms; gardens and lawns of private citizens; public sites such as golf courses, cemeteries, and roadways; land reclamation such as strip mines; and nonfood chain uses such as forest and grain for alcohol.

Table 2-4. Numerical metal limits for sludge application to cropland in the USEPA Part 503 sewage sludge regulations.

Metal	Ceiling limit†	Cumulative loading‡	Pollutant concentration limit§
	mg kg^{-1}	kg ha^{-1}	mg kg^{-1}
As	75	41	41
Cd	85	39	39
Cr	3000	3000	1200
Cu	4300	1500	1500
Pb	840	300	300
Hg	57	17	17
Mo	75	18	18
Ni	420	420	420
Se	100	100	36
Zn	7500	2800	2800

†The maximum allowable pollutant concentration in land applied sewage sludge.
‡The maximum amount of a pollutant that can be land applied if the pollutant concentration is below the ceiling, but above the pollutant concentration limit.
§Pollutant concentration in sewage sludge below which sewage sludge can be land applied without restrictive requirements and management practices.

The USDA and landgrant agricultural universities must also be credited for their research, and providing impetus for the progress in utilization of sludge on land. The USDA-ARS has had active involvement in conducting research related to land application of sewage sludge and its impact on soils, crops, animals, and water quality. The Cooperative Extension Service has provided guidance for implementing land application programs, and a forum for interacting with the public and farmers. The Soil Conservation Service has provided guidance for implementing land application programs by providing technical support and knowledge on soils, slopes, drainage and cropping practices. The USDA has also provided funds to agricultural experiment stations, and competitive grants to scientists for research.

Landgrant universities have played a major role in the land application of sewage sludge. Scientists at these locations have conducted research on the land application of sewage sludge and its impact on soils, crops, animals, and water quality. The scientists have also interacted with staff from the Cooperative Extension Service, the public, and farmers on the land application of sewage sludge.

Because of this research, the behavior of sludge-borne constituents in land application systems is much better understood today than it was 25 yr ago. Many USDA and university researchers have been outspoken proponents of land application, and they have tried to promote the practice based on technical information obtained from research and demonstration projects.

EXAMPLES OF SLUDGE APPLICATION TO LAND PROJECTS

During the past 25 yr, many municipalities have developed and implemented well-operated land application programs for sewage sludge. The largest of these programs was developed by the Metropolitan Water Reclamation District of Greater Chicago (District).

From 1970 to 1976, the District purchased 6289 hectares of land in Fulton County, IL, for the purpose of beneficially using sewage sludge to reclaim strip-mined lands. An extensive monitoring program for surface waters, ground waters, soils, and crops was set up, and this program is continuing. In-house and cooperative research has been conducted at the site to evaluate the impact of sludge application to land on animals, crops, soils, and water quality. Corn (*Zea mays* L.), winter wheat (*Triticum aestivum* L.) and forages are routinely grown on 1478 hectares of the site. Liquid sewage sludge at 4 to 6% solids was applied to application fields from 1972 to 1984. Dewatered sewage sludge at 60 to 70% solids has been applied from 1987 to the present. The District has reclaimed 140 ha of acidic coal refuse piles with sewage sludge from 1987 to 1993. A total of 1 063 000 dry metric tons of sewage sludge have been utilized at the site from 1972 to 1992. In 1991, the site received the Beneficial Use of Sewage Sludge Award from the USEPA.

Another municipality with a well-operated land application program for sewage sludge is the Madison Metropolitan Sewerage District in Wisconsin. In 1974, a decision was made to recycle sludge previously stored in lagoons to privately owned farmland as a fertilizer and soil conditioner. This District's long-term solids management program known as the Metrogro Program began in 1979. Liquid digested municipal sludge is applied by injection to area cropland growing mainly corn. The sludge is applied to 290 Department of Natural Resources (DNR) approved sites consisting of 12 150 ha within a 32 km radius of Madison, WI. Some income is generated by charging participating farmers $7.50 per acre for land receiving Metrogro applications. A monitoring program has been established for sludge, soil, plant tissue, and ground water. A total of 135 840 dry metric tons of sludge have been distributed from 1979 to 1992. The Metrogro Program has been used as a model by the USEPA and received an award as one of the most outstanding beneficial reuse programs in the USA (Taylor & Northouse, 1992).

Other local agencies have also developed successful land application programs for sewage sludge. The Milwaukee Metropolitan Sewerage District, besides distributing its nationally known heat-dried sludge called Milorganite®, is also marketing a liquid digested sludge called Agri-Life®. From 1975 to 1992, 247 025 dry metric tons of this sludge have been applied by injection to farmland in 10 counties near Milwaukee. The sludge is applied mainly to corn at rates consistent with crop need at sites inspected by the Wisconsin Department of Natural Resources (P.L. Schelect, Agri-Life Operations, Oak Creek, WI, 1993, personal communication). The City of Columbus, OH, has also developed a program for the beneficial use of sludge on land. An agronomist inspects all sites, and state permits are obtained. Sludge is applied at rates consistent with crop N requirements. A total of 83 685 dry metric tons of liquid digested sludge has been applied to farmland in the state of Ohio from 1972 to 1992 (J. Burrus, Dep. of Utilities, Solid Waste Reduction Facility, Columbus, OH, 1993, personal communication).

RESEARCH NEEDS

Research needs of benefit to municipal agencies in applying sewage sludge to land are as follows:

1. A second National Sewage Sludge Survey is needed to update the concentrations of constituents in sludge since the first survey conducted in 1987, and to determine what changes have occurred in sludge management practices.

2. The pathogen criteria in the Part 503 regulations are not based on risk-assessment methodology. Future pathogen criteria need to be based on sound risk-assessment methodology, and this methodology needs to be developed.

3. The plant availability of sludge-borne metals appears to be a function of metal species and other sludge, soil, and plant factors. Continuing research is needed to characterize these factors and to develop a model, such as that proposed by Dr. Richard Corey, to predict the plant uptake of sludge-borne metals.

4. Ecological research studies are needed to further refine the Part 503 regulations. Areas requiring additional research include:
   a. Bioavailability of sludge constituents to both plants and animals under different environmental conditions.
   b. Ecological effects of organic and inorganic constituents on wildlife and noncultivated crops, and impacts on unmanaged plant and animal communities.
   c. Long-term temporal changes at land application sites such as changes in soil binding capacity as sludge ages, and sensitivity of these temporal changes to changes in site condition.

CONCLUSIONS

Sludge application to land is becoming a major tool for managing municipal sludge in the USA. With the continued cooperation of government agencies and continuation of well-run sludge application to land projects, the trend toward increased use of this sludge management option will continue into the future.

REFERENCES

Council for Agricultural Science and Technology. 1976. Application of sewage sludge to cropland: Appraisal of potential hazards of the heavy metals to plants and animals. CAST Rep. 64. CAST, Ames, IA.

Council for Agricultural Science and Technology. 1980. Effects of sewage sludge on the cadmium and zinc content of crops. CAST Rep. 83, CAST, Ames, IA.

Logan, T.J., and R.L. Chaney. 1983. Utilization of municipal wastewater and sludge on land: Metals. p. 235–326. *In* A.L. Page et al. (ed.) Utilization of municipal wastewater and sludge on land. Denver, CO. 23-25 Feb. 1983. Univ. of California, Riverside.

National Association of State Universities and Land Grant Colleges. 1974. Proceedings of the Joint Conference on Recycling Municipal Sludges and Effluents on Land. Champaign, IL. 9–13 July 1973. NASULGC, Washington, DC.

Page, A.L., T.J. Logan, and J.A. Ryan (ed.) 1987. Land application of sludge: Food chain implications. Lewis Publ., Chelsea, MI.

Smith, J.E. 1977. Sludge treatment: Problems and solutions. Water Sewage Works 124:80–83.

Taylor, D.S., and M. Northouse. 1992. The Metrogro Program land application of anaerobically digested liquid sewage sludge at Madison, WI. Revised Feb. 1992. Madison Metropolitan Sewerage District, Madison, WI.

U. S. Department of Agriculture. 1989. Peer review, standards for the disposal of sewage sludge. USEPA proposed rule 40 CFR parts-257 and 503 USDA-CSRS Technical Committee. Univ. of California, Riverside.

U. S. Environmental Protection Agency. 1993. Standards for use or disposal of sewage sludge. Final rule, 40 CFR Part 503. Federal Register 58(32):9248–9415. 19 Feb. 1993. U.S. Gov. Print. Office, Washington, DC.

SECTION II

MANAGEMENT OF SOIL-PLANT SYSTEMS IN SEWAGE SLUDGE UTILIZATION

3 Interactions of Sewage Sludge with Soil–Crop–Water Systems

S. L. Oberle
D. R. Keeney

Department of Agronomy
Iowa State University
Ames, Iowa

The management and disposal of sewage sludge in an economical and environmentally acceptable manner is one of society's most pressing problems. Most municipal sludge is currently disposed of by landfilling, incineration, ocean dumping, and lagooning; however, land application of sludge for agricultural production and soil reclamation is becoming more popular. The growing preference for land application is due in part to economic, environmental, and regulatory constraints associated with alternative methods of sludge disposal (Elliott, 1986).

Although the benefits of using sewage sludge in agricultural production and land reclamation are well documented, land application of sludge can lead to potential problems. Constraints associated with use of sludge are generally related to sludge composition, site characteristics, management limitations, and public acceptance. Appropriate sludge utilization systems for agricultural production must ensure a proper balance between sludge applied nutrients and the nutrient needs of the particular crop being produced, and must consider potentially harmful effects such as the inducement of crop micronutrient deficiencies and phytotoxicity. Acceptable sludge management options must also minimize negative environmental impacts including degradation of the soil's capacity to sustain plant growth, and discharge of nutrients, especially nitrogen (N) and phosphorus (P), and other potentially harmful constituents to surface water and ground water.

CROP PRODUCTION

Sludges vary widely in their chemical, biological, and physical properties, depending on such factors as the source and composition of the sewage, the treatment system, the extent to which the material is digested and stabilized, and how the material is handled between processing and application to the soil. A

Sewage Sludge: Land Utilization and the Environment. SSSA Miscellaneous Publication.

major reason for considering the use of sewage sludge in crop production is its nutrient value. The guiding principles in designing an effective nutrient management program for use of sludge in cropping systems are to supply sludge-derived nutrients when and where they are needed by the crop, in the required quantities.

Sludge application rates in crop production are based on many factors, with primary considerations being the nutrient supplying capacity of the sludge material, the inherent soil fertility, the crop nutrient demand, and the sludge placement and timing. While many plant nutrients are present in sludge and numerous studies have shown crop responses to sludge additions, sewage sludges are generally considered a low-analysis nutrient source. Much effort has been directed toward assessing nutrient availability from various sludges and in general the nutrient supplying capacity of sludge has been very difficult to predict. This is in a large part due to differences in sludge quality as well as differences in the conditions under which the experiments were performed. Sludge nutrients, unlike those in commercial fertilizers, are not balanced to crop requirements (Elliott, 1986).

Differences in the nutrient availability from various sludges are generally related to differences in sludge composition and the particular soil-climate-management system in which the sludge is utilized. In order to ensure a well-balanced fertility program, sludge application rates should reflect the inherent nutrient supplying capacity of the soil, any nutrient additions from other sources (e.g., crop residues, animal manures, and commercial fertilizers), and the specific crop nutrient requirements. This helps ensure that sufficient nutrient levels are present to meet crop demand and that potentially toxic levels do not occur.

The timing of sludge application to cropland is a critical component of a safe and effective sludge management program. Sludge additions in crop production should be timed so that nutrient availability from the sludge coincides with the nutrient needs and uptake patterns of the crop. This aids in minimizing any negative environmental impacts associated with nutrient loss to surface water and ground water. Also, due to the high salt content of many sludges, crop growth can be inhibited if sludge is applied to soils at the wrong time (Keeney et al., 1975).

Sludge placement has been investigated by a number of researchers (Dunnigan & Dick, 1980; Bruggeman & Mostaghimi, 1993). Their findings indicate that surface application of sludge resulted in higher loading of N and P in runoff, as compared with areas in which the sludge was incorporated into the soil. Additional cropping systems components such as tillage can also have an impact on sludge management and utilization. Potential negative water quality impacts of sludge can be avoided by using sludge in conjunction with conservation practices, including tillage, that reduce erosion and prevent the movement of sediment and sludge constituents from the site of application to surface waters.

SOIL–WATER INTERACTIONS

In developing sludge management strategies for soil improvement it is useful to think of the soil as a medium for plant growth and other biological activity at or below the soil surface. Sludge additions to land will influence physical, chemical, and biological processes within the soil, and can lead to significant changes in soil–plant–water relations. Physical changes in soil associated with sludge application to land include enhanced soil organic matter and water holding capacity; as well as improved soil structure, aggregation, and water infiltration. This can lead to reduced runoff and erosion. For these reasons, sewage sludge has been used in many areas for the reclamation and revegetation of disturbed lands, and for improvement of marginal soils. Addition of sludge to marginal and disturbed soils can restore soil productivity and quality, and in general, provide a more favorable environment for plant growth. In many cases, however, appropriate sites are not available and sludge must be applied to high quality farmland, often surrounded by sensitive natural resource areas.

Chemical changes in soil resulting from sludge additions largely depend on the nature and amount of the chemical inputs from the sludge material and the assimilative capacity of the soil. Because of the extreme variability in sludges, it is critical to monitor the quality of sludge being considered for application to land, and to carefully manage sludge additions to control loading of nutrients and other chemical constituents such as metals, toxic organics, and salts. Failure to do so may degrade soil quality, restrict the capacity of the soil to support plant growth, and lead to entry of toxic metals in the food chain.

Biological considerations in land application of sludge are usually related to the presence of pathogens in the sludge. Because of the potential for animal and human health effects from improper land application, pathogen removal is necessary before sludges are suitable for agricultural use. The risk of exposure to pathogens can be further reduced through appropriate management measures including the use of buffer areas around application sites and placement of the sludge below the soil surface.

In the final analysis, there are many interacting social, economic, and environmental considerations that must be addressed in evaluating sludge use and disposal options, and in developing acceptable and effective sludge management practices. The relative social, economic, and environmental risks of all sludge management options must be considered in arriving at policies and practices to effectively manage sludge use on land. Awareness of the agricultural, health, and environmental benefits, costs, and impacts is critical to developing publicly acceptable sludge management and disposal programs that are both economical and effective in protecting the food chain and the environment (Elliott, 1986).

RESEARCH AND TECHNOLOGY NEEDS

Research and technology needs associated with land application of sewage sludge can be classified into two categories: component research and systems research. Component research, or that research that is required to fill existing knowledge gaps include, but are not limited to:

- Farmer attitudes and perceptions related to the risk of sludge use;
- Sludge management and behavior in reduced tillage systems;
- Economic costs and benefits to farmers and other sludge users;
- Assessment of sludge, soil, and tissue tests for sludge nutrient availability and toxicity; and
- Long-term nutrient availability, plant response, and environmental impact of using co-composted sewage sludge and municipal solid waste on agricultural lands.

In terms of systems research needs, future sludge management options on agricultural lands must be considered in both farming systems and agroecosystems contexts. Substantial progress can be made in sludge research and technology development simply by synthesizing existing information. From the perspective of farmers and land managers, application of sludge to agricultural lands can be an inequitable risk, where the principal risk-taker does not necessarily receive the major portion of the benefits (Elliott, 1986). Farmer and public acceptance of economic and environmental risks associated with utilization of sludge on agricultural lands must be considered in developing appropriate sludge management policies and practices.

Development of effective sludge management strategies also requires careful consideration of existing agricultural waste management issues. Meaningful solutions to waste related problems will be much simpler if practitioners and researchers educate one another and together work to solve the different problems of the producer (Schulte & Kroeker, 1976). One cannot expect that farmers with existing manure management problems due to excessive manure production, improper storage or application practices, or limited land resources will be willing to incorporate an effective sludge management program into their production systems.

REFERENCES

Bruggeman, A.C., and S. Mostaghimi. 1993. Sludge application effects on runoff, infiltration, and water quality. Water Res. Bull. 29:15–26.

Dunnigan, E.P., and R.P. Dick. 1980. Nutrient and coliform losses in runoff from fertilized and sewage sludge-treated soil. J. Environ. Qual. 9:243–250.

Elliott, H.A. 1986. Land application of municipal sewage sludge. J. Soil Water Conserv. 41:5–10.

Keeney, D.R., K.W. Lee, and L.M. Walsh. 1975. Guidelines for the application of wastewater sludge to agricultural land in Wisconsin. Wisconsin Dep. Nat. Resour. Tech. Bull. 88. Wisconsin Dep. Nat. Resour., Madison.

Schulte, D.D., and E.J. Kroeker. 1976. The role of systems analysis in the use of agricultural wastes. J. Environ. Qual. 5:221–226.

4 Plant Nutrient Aspects of Sewage Sludge

Gary M. Pierzynski

Department of Agronomy
Kansas State University
Manhattan, Kansas

The application of sewage sludge to agricultural land has been shown to be an agronomically and environmentally acceptable means of sludge disposal and for supplying plant nutrients. Approximately 35% of the publicly owned treatment works surveyed in the National Sewage Sludge Survey reported land application as the practice used to dispose of the majority of their sludge. The use of this disposal practice is likely to increase in the future as alternate practices become prohibitively expensive or are banned (e.g., landfilling, ocean disposal, or incineration). This chapter will discuss the advantages and disadvantages of using sewage sludges as sources of plant nutrients, with particular emphasis on nitrogen (N), phosphorus (P), and potassium (K), and the influence of sewage sludges on soil pH.

The nutrient contents of sludges vary considerably with ranges of <1 to 176 g kg^{-1} for total N, 5 to 67 600 mg kg^{-1} for NH_4–N, 2 to 4900 mg kg^{-1} for NO_3–N, <1 to 143 g kg^{-1} for total P, and 0.2 to 26.4 g kg^{-1} for K being reported (Sommers, 1977). Nutrient concentrations in sludges would not be expected to change much over time, unlike trace element concentrations that have declined. Two factors that could change nutrient concentrations would be the implementation of a phosphate ban in soaps and detergents on a statewide basis or the implementation of P removal practices by an individual treatment plant. In addition, sludge composition can vary considerably over time from a given treatment facility. Average coefficients of variation (expressed as a percentage of the mean) from a study of sludges from eight cities during 2 yr were 82% for inorganic N, 23% for organic N, 25% for inorganic P, 95% for organic P and 39% for K (Sommers et al., 1976). The intracity and intercity variability illustrates the need for sludge analyses on a frequent basis when the sludge is to be used in a land application program.

The recently passed standards for the use or disposal of sewage sludge (503 regulations) dictate that bulk sewage sludge cannot be applied at a rate that

Sewage Sludge: Land Utilization and the Environment. SSSA Miscellaneous Publication.

exceeds the agronomic rate. The agronomic rate provides the amount of N needed by the crop or vegetation grown on the land and minimizes the amount of N in the bulk sewage sludge that leaches below the root zone of the crop or vegetation to the ground water. Therefore, in the absence of some other limiting constituent in a sludge, careful consideration must be given to calculating the application rate based on information regarding the kinds and amounts of N in the sludge. A sludge analysis typically reports total N, NO_3–N (N_{NO_3}), and NH_4–N (N_{NH_4}). Organic N (N_{org}) is estimated as total N minus NO_3–N and NH_4–N. The N in sludge is or becomes available for the processes of nitrification, denitrification, immobilization, volatilization, and mineralization in the soil N cycle after the sludge has been applied. Nitrification is an aerobic process where NH_4–N is converted to NO_3–N. Denitrification is an anaerobic process where NO_3–N is converted to gaseous N forms (N_20 or N_2). Immobilization refers to the use of inorganic N forms by soil microorganisms in such a way that the inorganic N is no longer available for plant uptake. Volatilization is the loss of N as NH_3. Mineralization is the conversion of organic N to NH_4–N.

The goal in N management with sludge applications is to calculate the amount of N in the sludge that will be available to the subsequent crop after a sludge application. This value is sometimes called potentially available N (PAN). It is generally assumed that 100% of the NO_3–N is available for plant usage. Similarly, it is generally assumed that 100% of the NH_4–N is available for plant uptake unless the sludge is surface applied in which case an estimate of the amount of N volatilized can be made and subtracted from the PAN value. The difficulty is predicting the available N that will be mineralized, a value that is often called potentially mineralizable N (PMN). Various incubation procedures have been utilized with the overall goal of estimating mineralized N as a percentage of N_{org} in the sludge. The production of NO_3–N and NH_4–N over time or a measurement of CO_2 evolution over time are used as indicators of mineralization. Some attention has been given to developing a chemical extractant to use on sludges as a substitute for the incubation procedures to estimate PMN. None have gained wide-spread acceptance. For a given sludge, PMN depends on factors such as soil type, temperature, and soil moisture content. The value also depends on the type of sludge and the method of processing for the sludge.

Some representative values for PMN are 40% for sludge from a waste activation process, 30% for primary sludge, 30% for lime stabilized sludge, 25% for aerobically digested sludge, 15% for anaerobically digested sludge, and 5 to 10% for composted sludge. Using the information above, the concentration of PAN (typically expressed as percentage by weight) in a sludge can be estimated with:

$$PAN = N_{NO_3} + XN_{NH_4} + YN_{org}$$

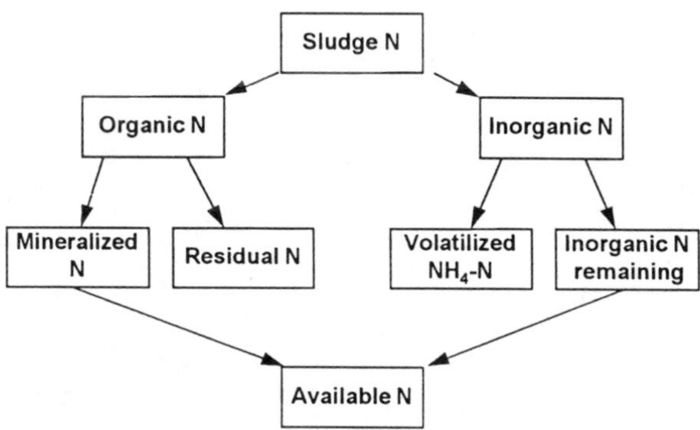

Fig. 4-1. Schematic of the procedure for estimating potentially mineralizable N (PAN) in sludges.

where X is fraction of NH_4–N that does not volatilize (often assumed to be 1) and Y is the fraction of N_{org} that is expected to mineralize. This concept is summarized in Fig. 4-1. Given the potential variability in sludge composition and in the values of X and Y, the most accurate application rates can be obtained through the use of frequent sludge analyses and incubation studies representative of each sludge disposal situation.

One potential disadvantage to supplying N for crops with sludge is the possibility of nitrate leaching, particularly after long-term sludge applications. This possibility exists because the mineralization process occurs whenever soil conditions are favorable, which, on an annual basis, occurs over a longer period than N uptake by the crop. Thus, inorganic N is produced in the soil when there is little chance of crop uptake. Any best management practice for maximizing the efficiency or minimizing the environmental impacts of inorganic fertilizer N would also be appropriate for sludge N (Keeney, 1982). Measuring residual soil NO_3 levels prior to a sludge application, and lowering the sludge application rate if warranted, would seem particularly relevant for sites receiving repeated sludge applications.

The P in sludges exists in both inorganic and organic forms with the inorganic forms generally predominating. Organic P must undergo mineralization in the soil before the P is available for plant uptake, similar to organic N, but the process has not received a great amount of research attention because the total P application rates are generally much higher than crop needs when a sludge application rate is based on PAN. For example, a sludge having 13 g kg^{-1} PAN and 10 g kg^{-1} total P applied to supply 150 kg PAN ha^{-1} would also apply 115 kg P ha^{-1}, which is approximately three times more than would

Table 4-1. The effect of two applications of lime-stabilized sludge on Bray-l extractable P concentrations and soil pH†.

Annual N rate			
Fertilizer	Sludge	Bray-l P	Soil pH
---------- kg ha^{-1} ------------		mg kg^{-1}	
0	0	35b‡	6.8c
67	0	22b	6.8c
200	0	30b	6.5c
0	67	40b	7.3b
0	134	43b	7.7a
0	200	68a	7.8a
0	267	92a	7.9a

†L.D. Maddux, Kansas State University, unpublished data, 1992.
‡Means followed by the same letter within a column are not significantly different (LSD, P=O.05).

typically be recommended for corn (*Zea mays* L.). Data in Table 4-1 shows the effects of two applications of a sludge on Bray-l extractable P levels. When 200 kg PAN ha^{-1} was supplied with the sludge for 2 yr, the Bray-l extractable P concentration approximately doubled compared with the control. The imbalance between N and P in sludges can cause soil P levels to increase substantially, often to levels much higher than necessary for adequate P nutrition of crops, and may increase the potential for off-site movement of P. Impairment of water quality because of excessive P levels has become a major concern for surface waters.

The Soil Conservation Service is in the process of adopting a P Site Index for their field office technical guide, developed to rank sites according to their potential to deliver P off-site, which has organic P application rate as one of eight site characteristics that contribute to this potential (Lemunyon & Gilbert, 1993). Phosphorus application rates as high as that in the example above could contribute significantly to the overall site vulnerability. It has been suggested that sludge application rates be based on the P content rather than on the N content. This would alleviate the increase in soil P that is common with long-term sludge applications, but would greatly increase the amount of land required to dispose of a given amount of sludge. Some states have compromised on this issue by requiring that sludge P applications not exceed anticipated crop removal of P after soil test P has reached a certain level (usually quite high). Phosphorus removal rates for most crops have been summarized (Pierzynski & Logan, 1993).

Sewage sludges are generally considered to be poor sources of plant available K, primarily due to the low concentrations of K in sludges. Potassium is a soluble constituent in sludges and when relatively high K concentrations

(>10 g kg^{-1}) are reported for sludges, this often reflects a sludge with a low solids content that has been dried down prior to analysis. The K in sludges, however, is normally assumed to be 100% available for plant uptake.

The addition of sewage sludge to soil can decrease, increase, or have no effect on soil pH. The mineralization of organic carbon (C) and N, nitrification of ammoniacal N forms, hydrolysis of iron (Fe) and aluminum (Al) compounds, and the oxidation of sulfides can all act to decrease soil pH. The addition of lime in the treatment process can act to increase soil pH. Land application of sludges with liming value in areas where acid soils are problematic can be beneficial. In areas where soil pH values are near or above neutral, such sludges can raise the pH to undesirable levels (Table 4-1). Sludge additions can also increase the pH buffering capacity of a soil.

There are several research needs that bear mentioning. There is a need for models or laboratory-based methods that can accurately predict PMN in sludges without the use of lengthy and laborious incubation procedures. Overall, this would allow a more accurate determination of application rates based on PAN and would minimize the chance of over or under applying N. A related research need is an evaluation of the use of soil profile nitrate testing prior to sludge applications. Sludge application rates could be reduced based on available N already present in the soil profile. This would reduce the potential for nitrate leaching, particularly on sites receiving sludges over many years. Finally, there is little information in the literature on soil test P as influenced by sludge applications when the application rates are based on the sludge P content, rather than the sludge N content.

In summary, sewage sludges can serve as acceptable sources of plant nutrients for crops with yields and crop composition comparing favorable with those produced through the use of commercial fertilizers. Proper management is required to minimize the potential environmental impacts.

REFERENCES

Keeney, D.R. 1982. Nitrogen management for maximum efficiency and minimum pollution. p. 605–649. *In* F.J. Stevenson (ed.) Nitrogen in agricultural soils. Agron. Monogr. 22. ASA, CSSA, and SSSA, Madison, WI.

Lemunyon, J.L., and R.G. Gilbert. 1993. The concept and need for a phosphorus assessment tool. J. Prod. Agric. 6:483–486.

Pierzynski, G.M., and T.J. Logan. 1993. Crop, soil, and management effects on phosphorus soil test levels. J. Prod. Agric. 6:513–520.

Sommers, L.E. 1977. Chemical composition of sewage sludges and analysis of their potential use as fertilizers. J. Environ. Qual. 6:225–232.

Sommers, L.E., D.W. Nelson, and K.J. Yost. 1976. Variable nature of chemical composition of sewage sludges. J. Environ. Qual. 5:303–306.

5 Trace Metal Movement: Soil–Plant Systems and Bioavailability of Biosolids–Applied Metals

Rufus L. Chaney

USDA-Agricultural Research Service
Beltsville, Maryland

Potential risk to feed- and food-chains from trace elements in biosolids (municipal sewage sludge) applied to agricultural land has been intensively examined during the last 25 yr. During this period, the *Soil–Plant Barrier* was described by which soil and plant chemistry prevent risk to animals from nearly all biosolids-applied trace elements mixed in soil. Adsorption or precipitation of metals in soils or in roots limits uptake-translocation to shoots of most elements, and phytotoxicity from zinc (Zn), copper (Cu), or nickel (Ni) limit residues of metals in plant shoots to levels chronically tolerated by livestock and humans. Another important understanding was that biosolids are not metal salts, but a mixture of metals that interact to reduce metal bioavailability, and adsorbing, chelating, and precipitating materials that also reduce phytoavailability and reduce bioavailability even when biosolids are ingested by grazing livestock or children. An example of the interaction phenomena is the toxicity of biosolids-applied Zn to animals; Cu-deficiency-stressed animals are more sensitive to dietary Zn than animals fed Cu-adequate diets, but biosolids-fertilized crops are not low in Cu. Similarly, Cu toxicity to sensitive ruminant animals is substantially reduced by increased dietary levels of Zn, cadmium (Cd), iron (Fe), molybdenum (Mo), and sulfate (SO_4); in contrast with the predicted toxicity from Cu in ingested biosolids, reduced Cu absorption occurs unless ingested biosolids exceed ≈1000 mg Cu kg^{-1}.

Another important interaction that reduces risk from biosolids-applied Cd is the normal 1:100 Cd/Zn ratio in biosolids with appropriate industrial pretreatment. Although culture of crops in strongly acidic soils allows uptake of increased levels of Cd and Zn, the presence of Zn in the crops reduces the potential for risk from Cd. Zinc phytotoxicity serves as a natural maximal limit on crop Cd, and plant (intrinsic) Zn inhibits absorption of plant Cd in animals (McKenna et al., 1992). These factors very significantly reduce biosolid/soil Cd

---

Sewage Sludge: Land Utilization and the Environment. SSSA Miscellaneous Publication.

risk compared with that estimated by toxicological studies in which risks from Cd salts were tested.

The potential for plant uptake to allow biosolid/soil trace elements to be transferred to the feed- and food-chains has been extensively studied. Uptake slopes measured in pot studies are much higher than found in the field, so greenhouse or growth chamber studies cannot be used to estimate uptake slopes. Unfortunately, only a small fraction of the valid field studies were conducted under poor management conditions rather than recommended *best management practices* for biosolids utilization. It is very clear that strongly acidic soils increase plant uptake of Zn, Cd, Ni, manganese (Mn), and cobalt (Co), and increase the potential for phytotoxicity from Cu, Zn, and Ni; alkaline soil pH increases uptake of Mo and selenium (Se), while biosolids applied lead (Pb) and chromium (Cr) are not absorbed to any significant extent at any pH (Chaney & Ryan, 1993). For the strongly adsorbed metal cations, the pattern of response has been found to be strongly curvilinear [plant metal concentration approaches a plateau with increasing biosolids application rate (Chaney & Ryan, 1993), at least for high quality biosolids] rather than a linear plant/soil relationship with increasing biosolids application rate if several potential errors in the research methodology are avoided. First, comparison of application rates is only valid after the period of rapid biodegradation of biosolids-applied organic matter; when high biosolids rates are applied, uptake can be increased for several years due to formation of biodegradation by-products that increase metal diffusion and convection to the roots. The effect is more significant for *Poaceae* than other species, perhaps due to the role of phytosiderophores in metal uptake. Second, soil pH levels should be equal across rates studied, or co-variance of soil pH used to correct for unequal soil pH. Studies by Bell et al. (1988) found a strong plateau response of plant Cd, Cu and Zn to biosolids-applied metals when pH co-variance was incorporated in the data assessment. Third, the metal concentration in the biosolid affects the slope of metal uptake (or the increase above the background plant metals when the plateau is reached); higher biosolid metal concentration means higher phytoavailability at equal metal applications (Jing & Logan, 1992). The presence of metal adsorbing sesquioxides in the biosolid decreases the slope or increment at the plateau.

It is now clear that, for most trace elements in biosolids, biosolids from treatment works with appropriate industrial pretreatment cannot cause adverse effects on humans, livestock, plants, or wildlife when >1000 Mg ha^{-1} is applied. The natural limitations on metal uptake and bioavailability, including the plateau response, prevent toxic levels being reached in plants used as food or feed. Several elements are normally low in biosolids (not discharged or not collected during sewage treatment), and only rare biosolids comprise risk. Regular analysis of biosolids provides adequate information to prevent adverse effects of biosolids trace elements on feed- and food-chains. Occasional biosolids contain such high levels of Cd, Mo, Zn, Pb, Cu or polychlorinated biphenyls (Chaney et al., 1991) that they should not be used on land. The new Clean Water Act 503 Rule uses this approach of the Alternative Pollutant Limit (APL) for higher quality biosolids. Significant errors were made, however, during the finalization of the 503 Rule. The 503 Cd limit (39 kg ha^{-1}) is based on ingestion of biosolids by children (Pathway 3), assuming 100% bioavailability of Cd in biosolids; research has clearly shown the bioavailability is much lower and this Pathway does not limit beneficial use of biosolids.

I conclude that the calculation of Cd limits for the garden foods Pathway (2) was in error at 120 kg Cd ha^{-1}; if arithmetic mean slopes for leafy vegetable Cd uptake from acidic soils were used rather than geometric mean slopes for all pH levels (only 21% of field slopes were for pH < 6.0), the APL Cd limit would have been ≈12 mg kg^{-1}, not 120 mg kg^{-1} estimated by USEPA for this pathway. By omitting slopes for field studies that used highly Cd contaminated biosolids (>150 mg Cd kg^{-1}), the APL Cd limit was estimated to be 21 mg kg^{-1}. Further, the pathways did not include consideration of whether crops could be marketed within Cd limits established by several nations. The USA has no crop Cd limits, largely because no Cd risk has been identified from normal U.S. crops. Several countries, however, have set grain Cd limits of 0.10 mg Cd kg^{-1}. Even with biosolids containing 21 kg Cd ha^{-1}, wheat (*Triticum aestivum* L.), soybean [*Glycine max* (L.) Merr.], peanut (*Arachis hypogaea* L.), and sunflower (*Helianthus annuus* L.) grain exceeds 0.1 mg Cd kg^{-1} when soils are strongly acidic. The USEPA has agreed to cooperate with USDA to develop guidance for application of biosolids on land that may be used to produce crops for export, and that would need to meet stricter Cd limits than in the USA.

Protection of wildlife is similar to that of livestock, however, wildlife may consume 100% of their diet from plants grown on biosolids-amended soils under worst case management conditions. In the case of wildlife in nonmanaged ecosystems, maximal plant residues may exceed those reached on managed farmland. Evaluation of a rich literature on wildlife exposure to metal contaminated soils indicates that animals that consume earthworms comprise the highly exposed individuals. Earthworms contain 45% soil on a dry matter basis. The soil + biosolids in the earthworm, however, can adsorb metals, reducing the bioavailability of the soil/biosolid metals. At the levels allowed under the Clean Water Act-503 (21 kg Cd ha^{-1} according to the USDA revision of the USEPA limit; 300 mg Pb kg^{-1}), body burdens in wildlife are far below toxic levels. Of the trace elements in biosolids, only mercury (Hg) in the methyl-Hg form can actually be biomagnified, but very little of the Hg in biosolids is in this form. Other metals are instead *biominified* according to many studies (Nelson et al., 1993). Nearly all of the metals in forage materials ingested by earthworms are excreted, in some case >99%. Thus, the increase in risk with increasing trophic level seen with chlorinated hydrocarbons has not been seen with biosolids-applied metals.

Because limed-biosolids caused Mn deficiency in Maryland and Virginia Coastal Plain soils, which had been depleted of soil Mn during soil genesis, we have conducted experiments using additions of Fe and Mn to biosolids compost to evaluate the potential of added Fe or Mn to reduce compost or soil Cd phytoavailability and prevent lime-induced Mn deficiency. Added Fe or Mn reduced uptake of Cd and Zn by lettuce (*Lactuca sativa* L.) somewhat, but Zn in biosolids was not as phytoavailable to reduce Cd uptake by lettuce as Zn added as $ZnSO_4$.

In comparison with the 1973 meeting on agricultural use of sewage sludge at the University of Illinois, we have come a long way in understanding the fate and potential effects of biosolids-applied metals and organics. Both practical and basic understandings have improved. Excessive concerns about metal phytotoxicity and Cd uptake are now known to have resulted from incomplete understandings of metal adsorption, metal bioavailability, and metal interactions. It now seems clear that high quality biosolids and similar materials may be used beneficially in sustainable agriculture. Recovery of the fertilizer and soil conditioner benefit, and reduced cost of ultimate disposal, by use of these materials on cropland does not comprise risk to humans, livestock, or wildlife.

IDENTIFIED RESEARCH NEEDS

1. The bioavailability of trace elements in plant shoots (intrinsic metals) at concentrations that cause significant yield reductions (e.g., >400 mg Zn kg^{-1}, >40 mg Cu kg^{-1}, or >50 mg Ni kg^{-1}) needs to be better characterized (compared with metal salts added to practical diets) so that the relative effectiveness of plant absorbed Cd, Zn, Cu, Mo, and Se in biosolid-amended soils are available for the pathway risk assessment. The effect of dietary deficiency of interacting metals should be evaluated, including the ability of plant absorbed trace elements to correct those deficiencies and reduce the potential for risk.

2. The persistence of the biosolids-applied specific metal adsorption capacity should be better characterized so that *designer biosolids* with inherently lower metal bioavailability and phytoavailability can be manufactured. The ability of Fe, Mn, or aluminum (Al) additions during sewage treatment or biosolids processing to increase the specific metal adsorption capacity of biosolids should be more fully characterized. The persistence of high adsorbing forms of the metal oxides in biosolids-amended soils needs to characterized further, and methods to stabilize metal adsorption capacity of biosolids identified.

3. The one food-web biomagnification concern among trace metals, CH_3-Hg, should be evaluated under the worst case condition in which bottom sediments are comprised of eroded biosolids residues, and subsistence fisherpersons consume fish which ingest sediments or macrofauna that accumulate CH_3-Hg from the sediment. The extent of formation of CH_3-Hg under these conditions appears to be small from the limited research conducted, but further research is needed. Similarly, mushrooms accumulate mostly inorganic Hg from compost, and the limits of compost-Hg needed to produce marketable mushrooms with acceptable Hg residues should be clarified.

4. The bioavailability of metals in biosolids and soil-biosolids mixtures ingested by livestock and wildlife should be more fully characterized (Freeman et al., 1992). This includes the case in which earthworm-consuming mammals and birds ingest high amounts of soil/biosolids mixtures. The maximum safe *soil* concentration for this pathway should be established for all trace metals.

REFERENCES

Bell, P.F., C.A. Adamu, C.L. Mulchi, M. McIntosh, and R.L. Chaney. 1988. Residual effects of land applied municipal sludge on tobacco: I. Effects on heavy metals concentrations in soils and plants. Tobacco Sci. 32:33–38.

Chaney, R.L., and J.A. Ryan. 1993. Heavy metals and toxic organic pollutants in MSW-composts: Research results on phytoavailability, bioavailability, etc. p. 451–506. *In* H.A.J. Hoitink and H.M. Keener (ed.) Science and engineering of composting: Design, environmental, microbiological and utilization aspects. Ohio State Univ., Columbus.

Chaney, R.L., J.A. Ryan and G.A. O'Connor. 1991. Risk assessment for organic micropollutants: U.S. point of view. p. 141–158. *In* P. L'Hermite (ed.) Treatment and use of sewage sludge and liquid agricultural wastes. Elsevier Applied Science, New York.

Freeman, G.B., J.D. Johnson, J.M. Killinger, S.C. Liao, P.I. Feder, A.O. Davis, M.V. Ruby, R.L. Chaney, S.C. Lovre, and P.D. Bergstrom. 1992. Relative bioavailability of lead from mining waste soil in rats. Fundam. Appl. Toxicol. 19:388–398.

Jing, J., and T.J. Logan. 1992. Effects of sewage sludge cadmium concentration on chemical extractability and plant uptake. J. Environ. Qual. 21:73–81.

McKenna, I.M., R.L. Chaney, S.H. Tao, R.M. Leach, Jr., and F.M. Williams. 1992. Interactions of plant zinc and plant species on the bioavailability of plant cadmium to Japanese quail fed lettuce and spinach. Environ. Res. 57:73–87.

Nelson, W., W.N. Beyer, and C. Stafford. 1993. Survey and evaluation of contaminants in earthworms and in soils derived from dredged material at confined disposal facilities in the Great Lakes. Environ. Monitor. Assess. 24:151–165.

6 Sewage Sludge: Toxic Organic Considerations

G. A. O'Connor

Department of Soil and Water Science
University of Florida
Gainesville, Florida

Numerous man-made organic chemicals with a wide range of chemical properties can occur in sewage sludges. One general grouping of these compounds is regarded as priority pollutants (referred to herein as TOs), and typically occurs in sludges in the milligram per kilogram concentration range. Concerns about the environmental fate of TOs threatened routine land application of sludge. A review of the pertinent literature (e.g., field or pot studies utilizing *sludge-borne* TOs, and appropriate analytical techniques) along with risk assessment, suggest that the concern is largely groundless.

The vast majority of sludge-borne TOs occur at low concentrations that are reduced at least 100-fold in typical (agricultural) land application scenarios. Further, most TOs are so strongly sorbed in the sludge-soil matrix as to have low bioavailabilities to plants, and are accumulated, if at all, at very low concentrations in the edible portion of food chain crops (O'Connor et al., 1991). Field studies confirm the negligible bioavailability to plants of sludge-borne TOs.

Much more important pathways of TO (especially persistent lipophilic organics) impact on the environment are: i) direct ingestion of sludge by children, ii) human consumption of meat from animals grazing forage surface-treated with sludge, iii) predators consuming biota living in sludge-amended soil, and iv) humans living in sludged-areas and exposed to air or drinking water and fish coming from surface sources.

The U.S. Environmental Protection Agency (USEPA) recently completed a detailed risk assessment for these and other potential pathways of environmental exposure. The assessment confirmed the expected low bioavailability of TOs to plants, animals, and humans — especially at the sludge loading rates typical of land application (Chaney et al., 1991). These findings, along with other considerations, result in TOs being unregulated in the recently promulgated 503 sludge rule, with regard to land application.

Sewage Sludge: Land Utilization and the Environment. SSSA Miscellaneous Publication.

Round two of the rule development, however, will revisit risk assessment of polychlorinated biphenyls (PCBs) and other organics deserving first-time consideration. Two such groups of compounds deserving round two consideration are: i) chlorinated dibenzo-*p*-dioxins (CDDs) and dibenzo furans (CDFs), and ii) aromatic surfactants (e.g., linear alkylbenze sulphonates and ethoxylates). The former chemicals typically occur in sludges at very low (microgram per kilogram) concentrations, but carry significant notoriety as extremely toxic compounds. Careful research must be conducted now to foster sound regulatory decisions in which policy concerns are clearly distinguished from scientific concerns.

The second group of chemicals is indicative of anthropogenic (and perhaps natural) organics that can occur in sludges at gram per kilogram concentrations. These much greater initial concentrations could greatly increase plant, animal, and environmental exposure, even at the modest sludge loading rates typical of land application. Little is known about the environmental behavior or toxicity of many such compounds, particularly in sludge-amended soils.

More research is also needed on the environmental behavior (process-level) of even well known TOs representing broad chemical groups, e.g., volatiles, polycyclic aromatic hydrocarbons (PAHs), and PCBs. USEPA's risk-assessment methodology required data (e.g., bioaccumulation values, sorption–desorption coefficients, and model components) that were either poorly understood or scarcely characterized. Extending the assessment process appropriately to new chemicals or new situations, which is likely to occur, necessitates more and better data to accurately assess risk. The fate and transport portion of the current methodology is particularly simplistic in its treatment of soil TO interaction, and deserves careful scrutiny and, possibly, reformulation.

REFERENCES

Chaney, R.L., J.A. Ryan, and G.A. O'Connor. 1991. Risk assessment for organic micropollutants: U.S. point of view. p. 141–158. *In* P. L'Hermite (ed.) Treatment and use of sewage sludge and liquid agricultural wastes. Elsevier Applied Sci., New York.

O'Connor, G.A., R.L. Chaney, and J.A. Ryan. 1991. Bioavailability to plants of sludge-borne toxic organics. Rev. Environ. Contam. Toxicol. 121:129–155.

7 Sewage Sludge: Pathogenic Considerations

J. Scott Angle

Department of Agronomy
University of Maryland
College Park, Maryland

Land application of sewage sludge requires that the risks associated with this practice be addressed during initial permitting. One risk that has received extensive study is the presence of pathogens in sludge and the potential for survival of pathogens following land application. Movement of pathogens from soil into surface and ground water resources has also received extensive consideration.

Wastewater entering a sewage treatment plant contains a wide variety of pathogenic organisms including bacteria, viruses, fungi, and parasites (protozoa and helminths). Organisms responsible for diseases such as Hepatitis A (Hepatitis A virus), AIDS (HIV), gastroenteris (*Escherichia*, *Salmonella*, and *Shigella*), cholera (*Vibrio cholerae*), and giardiasis (*Giardia lambia*) have all been isolated from raw wastewater and sludge (Yanos, 1987; Goldstein et al., 1988). The fungus, *Aspergillus fumigatus*, that causes a pulmonary disease referred to as *farmers lung* has also been isolated from sludge during composting. This is the only known fungal pathogen associated with sludge. This organism is typically only a problem for individuals with a compromised immune system (Burge & Miller, 1980). Since it is both expensive and dangerous to work with pathogenic organisms, many monitoring programs assay for *indicator organisms* in sludge. Total coliforms, fecal coliforms and fecal streptococci have been used for many years as indices of human contamination of soil and water (American Public Health Association, 1985).

One of the primary functions of sewage treatment is to eliminate or reduce the pathogen load in sludge and wastewater. The extent of reduction is dependent upon the type of treatment applied to the sludge and the organisms that are present in the sludge. Primary treatment significantly reduces the number of pathogenic and indicator bacteria in sludge compared with numbers present in wastewater. Reductions of >90% are often observed following primary treatment. Viruses tend to be relatively stable during sewage treatment

Sewage Sludge: Land Utilization and the Environment. SSSA Miscellaneous Publication.

with reductions of ≈50% during primary treatment and only slightly greater during secondary treatment (Farrah & Bitton, 1990). The same is true for parasites, with many surviving primary and secondary treatment. Parasites often form environmentally resistant structures that withstand degradation during processing. Pathogenic bacteria, on the other hand, tend to be environmentally sensitive and most bacteria are killed during primary treatment. It is only through tertiary treatment that many of the viruses and parasites found in sludge are killed (Farrell & Stern, 1975).

The U.S. Environmental Protection Agency has issued standards (40 CFR Parts 257, 403, and 503) for land application of sludge based upon the number of pathogens and indicator organisms in sludge. Class A sludge can be used for any purpose and is defined as sludge that has <1000 fecal coliform bacteria per gram, less than three *Salmonella* sp. per four grams, less than one helminth ova per four grams, and less than one enteric virus per four grams. Class B sludge contains significantly higher numbers of pathogens and indicator bacteria. This sludge can be applied to land, however, use is subject to numerous site restrictions. For example, root crops grown in soil amended with Class B sludge cannot be harvested for up to 38 mo following land application. Public access to land amended with class B sludge is also limited for a period of up to one year. To achieve Class A status, it is frequently necessary to subject the sludge to tertiary treatment. This could include processes such as composting, heat drying, heat treatment, alkaline stabilization, thermophilic digestion, beta and gamma ray irradiation, and pasteurization.

Organisms applied to soil via land application are subject to a variety of fates. Of greatest concern is the potential to survive in soil for extended periods of time. Chemical and physical properties of soil significantly affect the survival of pathogens added to soil during land application. Most pathogens rapidly die in hot, dry soils and survive for long periods of time in cold, moist soils (Sagik & Sober, 1978). Low soil pH and low organic matter content also favor rapid death of pathogens in soil (Gerba et al., 1975). Injection or rapidly plowing sludge into soil enhances the survival of pathogens in soils since sunlight, and especially ultraviolet radiation, kills many types of pathogens.

Representative survival times for important groups of pathogens are summarized in Table 7-1. Bacterial pathogens typically die relatively quickly in soil, with most pathogens failing to survive beyond a few weeks. Viruses survive for longer periods of time since they exhibit few metabolic functions. Enteroviruses may survive for up to 6 mo in soil. The survival of viruses in soil has been reviewed in detail by Farrah and Bitton (1990). The Hepatitis A virus has been shown to persist in soil longer than many other viruses and may partially explain why waterborne outbreaks of this disease have been traced to consumption of contaminated ground water (Sobsey et al., 1986). Parasites form environmentally resistant structures such as cysts or ova and can survive for years in soil with some organisms reported to survive for up to 10 yr (Bergstrom & Langeland, 1981).

Extended survival in soil suggests that several other environmental fates may potentially affect human health. Pathogenic organisms, due to their small size, can potentially leach through the soil profile and into ground water (Gerba

Table 7-1. Survival times of pathogens in soil.†

Organism	Approximate survival time
	--- d ---
Fecal coliforms	8–55
Leptospira	15
Streptococcus faecalis	26–77
Enterovirus	70–170
Poliovirus	70–90
Helminth eggs	>1000

†Adapted from Kowal (1986).

et al., 1975). Viruses have been shown to leach the greatest distance through soil as a result of their small size and relatively weak interaction with clay and organic materials. Viruses have been reported to leach many meters into soil before contacting ground water. Bacteria leach to a more limited extent compared with viruses. Bacteria are larger and thus subject to physical screening by soil particles. Interactions between soil particles and the cell surface further prevents bacteria from leaching through soil. Parasites, especially large ova and cysts, seldom leach into soil to any significant extent. These organisms are retained near the point of application or injection by physical screening filtering.

Chemical and physical properties of soil have a significant effect on leaching. Pathogens leach much further into gravelly soils of low exchange capacity compared to fine textured soils of high exchange capacity. Other factors that significantly affect leaching include soil pH, soil texture, water content, and the presence of plant roots (Berry & Hagedorn, 1990; Gammack et al., 1992). Leaching of pathogens is greatest under conditions of saturated flow (Lance & Gerba, 1984). Hagedorn et al. (1981) showed that when bacteria are forced through soil under saturated flow, the majority of movement is through macropores where interaction with soil colloids and particulates are limited. Saturated flow is important during direct application of wastewater to land and septage flow from tile lines.

Pathogens that remain at or near the soil surface are subject to runoff losses into surface water. Detachment from soil particles or runoff of sediment-associated pathogens have been shown to occur during heavy rainstorm events. Although few studies have actually measured losses from sludge-amended land, many studies have examined runoff losses from soil amended with manure. Most studies have reported that losses via runoff are minimal. Since runoff losses occur only from the top few millimeters of soil, movement is minimized since few pathogens survive directly on the soil surface. Ultraviolet radiation rapidly kills most pathogens exposed to sunlight. Only when an intense storm occurs soon after sludge application, is runoff a significant problem. Incorporation of sludge into soil effectively eliminates runoff as a potential concern.

There are few pathogenic hazards posed by the land application of sludge, especially in comparison to other materials in sludge such as heavy metals, toxic organisms, or soluble salts. Ottolenghi and Hamparian (1987) reported that there was no increase in serum levels of antibodies to *Salmonella* sp. obtained from families living on farms where sludge was applied to soil. Although there is little presumption that sludge-borne pathogens are a significant health concern, several important questions remain. As land application increases, it will become necessary to utilize models that will allow one to predict survival and movement (Corman et al., 1982). Currently, there are no adequate models that accurately predict survival in soil. Since survival is closely related to the potential for runoff and leaching, most transport models are also lacking in precision due to the omission of these parameters. Soil specific input data (pH, cation–exchange capacity, texture, and organic matter) will also be required if accurate models are to be developed.

Another important area that requires continuing attention is the presence of new pathogens in sludge. For example, questions have only recently been raised as to whether the virus that causes AIDS can leach from sludge-amended soil into ground water. Ansari et al. (1992) has recently shown that HIV may survive for short periods of time in wastewater and Yolken et al. (1991) has suggested that fecal shedding may play a role in transmission within the environment. A further area that will require increased study during the next decade are methods to prevent pathogens from moving from sludge-amended land into surface and ground water. Numerous methods have recently been developed for the prevention of nutrient and pesticide losses from amended fields. It is important to determine whether these, or other best management practices, can prevent movement of sludge-borne pathogens.

REFERENCES

American Public Health Association. 1985. Standard methods for the examination of water and wastewater. APHA, Washington, DC.

Ansari, S.A., S.R. Farrah, and G.R. Chaudhry. 1992. Presence of human immunodeficiency virus nucleic acids in wastewater and their detection by polymerase chain reaction. Appl. Environ. Microbiol. 58:3984–3990.

Bergstrom, K., and G. Langeland. 1981. Survival of ascarsis egg, salmonella, and fecal coliforms in soil and on vegetables grown in infected soil. Nord. Veterinaermed. 33:23–32.

Berry, D.F., and C. Hagedorn. 1990. Soil and groundwater transport of microorganisms. p. 57–73. *In* L.R. Ginzburg (ed.) Assessing ecological risks of biotechnology. Butterworth-Heinemann Publ., Boston.

Burge, W.D., and P.B. Millner. 1980. Health aspects of composting: Primary and secondary pathogens. p. 245–264. *In* G. Bitton et al. (ed.) Sludge-health risks of land application. Ann Arbor Sci. Publ., Ann Arbor, MI.

Corman, A., Y. Crozat, and J.C. Cleyet-Marel. 1982. Modelling of survival kinetics of some *Bradyrhizobium japonicum* strains in soil. Biol. Fertil. Soils 4:79–84.

Farrah, S.R., and G. Bitton. 1990. Viruses in the soil environment. p. 529–551. *In* J.M. Bollag and G. Stotzky (ed.) Soil biochemistry. Vol. 6. Marcel Dekker, New York.

Farrell, J.B., and G. Stern. 1975. Methods for reducing the infection hazard of wastewater sludge. p. 21–29. *In* Radiation for a clean environment. Int. Atomic Energy Agency, Munich.

Gammack, S.M., E. Peterson, J.S. Kemp, M.S. Cresser, and K. Killham. 1992. Factors affecting the movement of microorganisms in soils. p. 263–290. *In* G. Stotzky and J.M. Bollag (ed.) Soil biochemistry. Vol. 7. Marcel Dekker, New York.

Gerba, C.P., C. Wallis, and J.L. Melnick. 1975. Fate of wastewater bacteria and viruses in soil. J. Irrigation Div. ASCE 101-IR3. ASCE, Washington, DC.

Goldstein, N., W.A. Yanko, J.M. Walker, and W. Jakubowski. 1988. Determining pathogen levels in sludge products. Biocycle 29:44–47.

Hagedorn, C., E.L. McCoy, and T.M. Rahe. 1981. The potential for groundwater contamination from septic effluents. J. Environ. Qual. 10:1–8.

Kowal, N.E. 1986. Health considerations in applying minimum treated wastewater to land. p. 27–54. *In* Utilization, treatment and disposal of waste on land. SSSA, Madison, WI.

Lance, J.C., and C.P. Gerba. 1984. Virus movement in soil during saturated and unsaturated flow. Appl. Environ. Microbiol. 47:335–337.

Ottolenghi, A.C., and V.V. Hamparian. 1987. Multi-year study of sludge application to farmland: Prevalence of bacterial enteric pathogens and antibody status of farm families. Appl. Environ. Microbiol. 53:1118–1124.

Sagik, B.P., and C.A. Sober. 1978. Assessing risk of effluent land application. Water Sewage Works 125:40–42.

Sobsey, M.D., P.A. Shields, F.H. Hauchman, R.L. Hazard, and L.W. Caton, III. 1986. Survival and transport of hepatitis A virus in soils, groundwater and wastewater. Water Sci. Technol. 10:97–106.

Yanos, T.M. 1987. Land application of wastewater sludge: A report of the task committee on land application of sludge. ASCE, Washington, DC.

Yolken, R.H., S. Li, J. Perman, and R. Viscidi. 1991. Persistent diarrhea and fecal shedding of retroviral nucleic acids in children infected with human immunodeficiency virus. J. Infect. Dis. 164:61–66.

8 Engineering and Cost Considerations: Sludge Management and Land Application

William J. Jewell

Department of Agricultural and Biological Engineering
Cornell University
Ithaca, New York

More than two billion dollars are spent annually treating and disposing of nearly 5.4 million dry tons of municipal sewage sludge in the USA. Septage generated in the rural areas is estimated to be ≈8.6 billion gallons (1.3 million dry tons per year) per year and 24% of the dry matter contained in centralized generated sewage sludge (U.S. Environmental Protection Agency, 1993).

Promulgation of new sludge management regulations has raised interest in defining sludge management options and their costs. The purpose of this chapter is to provide a brief overview of engineering and cost considerations relating to land application of sludge and to review several promising technologies. The U.S. Environmental Protection Agency (USEPA) waste management policy promotes reduction and reuse of waste. For this reason destruction technologies, such as incineration, will not be considered here.

SLUDGE GENERATION, TREATMENT, AND DISPOSAL

Defining a Quantitative Frame of Reference

It is important to have a quantitative understanding of sludge solids generated and technologies available to reduce problems that these materials create and the economics associated with these options. Manuals developed by USEPA provide guidance on 18 major processes with 44 variations (U.S. Environmental Protection Agency, 1985), and land application options provide evaluation of 16 different variables (U.S. Environmental Protection Agency, 1983). Many uncertainties exist, however, relating to quantification of sludge treatment and disposal problems. Example areas that need clarification include:

Sewage Sludge: Land Utilization and the Environment. SSSA Miscellaneous Publication.

- good definition of expected *average* domestic sludge generation rates from conventional sewage treatment plants;
- sludge volume and mass reductions achieved with conventional technologies;
- new and innovative technologies available that would reduce this amount;
- conventional alternative cost-effectiveness; and
- cost reduction potential of new technologies.

A quantitative frame of reference is essential to measure the effectiveness of specific installations as well as to provide guidance on future needs and potentials. Reported sludge generation rates are equivalent to an average production of 21.4 kg (dry) per person per year (47 lb per person per year). Since the theoretical average sewage solids production is ≈34.5 kg per person per year, >60% of the solids produced by the population requires final disposal from conventional sewage plants.

Septage results in the generation of 14.5 kg per person per year, and this represents a final disposal problem that represents ≈42% of the dry matter generated on average. The large quantities of sludge requiring final disposal is disappointing considering the high cost and the technologies available to reduce sludge volumes and mass.

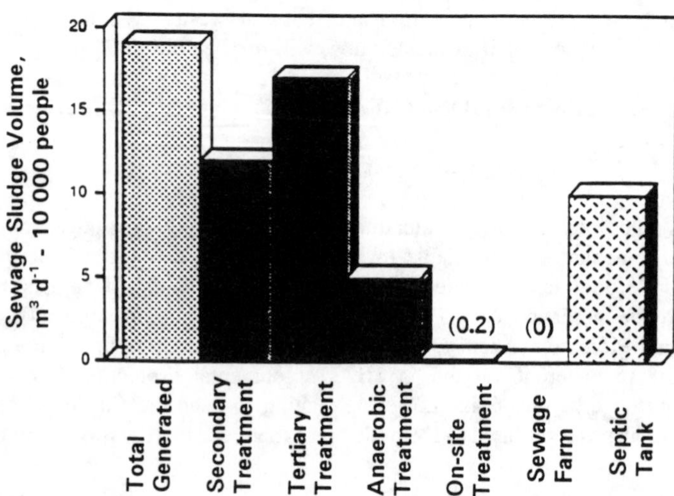

Fig. 8-1. Comparison of wet sludge generated from various treatment alternatives for typical domestic sewage.

A perspective comparing the total amount of wet sludge generated by domestic populations and various alternative treatment systems is shown in Fig. 8-1. Total solids generated in wastewater would result in the production of ≈19 m^3 per 10 000 people per day. The fate of the wet and dry mass in a typical secondary biological treatment plant are illustrated in Fig. 8-2. This shows that the applications of several separation and stabilization processes reduce sludge volumes by only 34%. Tertiary treatment plants can generate more sludge. It is interesting to note that if an average residence with 3.5 people generates a septage volume of 1000 gallons every 3 yr with a 4% dry matter content, it represents approximately the same volume of sludge as a secondary biological treatment plant.

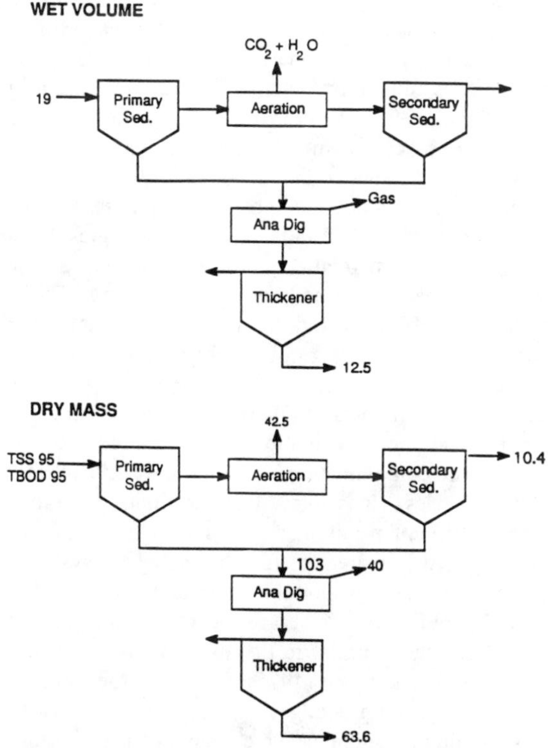

Fig. 8-2. Overview of fate of sludge in a typical secondary aerobic biological domestic sewage treatment facility. Values in g (dry) per person per day or m^3 per day per 10 000 people (wet volume).

INNOVATIVE SLUDGE MINIMIZATION TECHNOLOGIES

Technologies exist or are under development that can minimize sludge production. Anaerobic biological treatment of wastewater can achieve significant reduction of sludge since the anaerobic biological system generates 75% less secondary sludge than aerobic treatment alternatives because of the different microbial yield under aerobic and anaerobic conditions. Anaerobic treatment has not been applied to domestic sewage because the slow-growing methanogenic organisms were not thought to be efficient enough to achieve good effluent quality (Jewell, 1988). Recently a 4-yr pilot study has been completed at Cornell University that demonstrated that the new attached microbial film expanded bed process is efficient enough to be considered as a secondary treatment system with minimum additions (Jewell et al., 1993). This study tested wastewater flows up to 40 m^3 d^{-1} (10 000 gallons per day) and used raw sewage fed directly to the anaerobic bioreactor. By allowing the raw sludge to become stabilized in the anaerobic wastewater system, it was hypothesized that a minimum quantity of sludge would be generated by this system.

Total mass balances in the anaerobic treatment of domestic sewage showed that the average sludge production was equal to 3.2 m^3 per day per 10 000 people. This results in a 65% sludge volume reduction over a conventional aerobic biological treatment plant. This reduction would appear to be close to the minimum achievable in a secondary quality effluent system.

Further reduction of sludge is possible using on-site treatment and dedicated natural systems (Jewell, 1980). Two technologies are in their early stages of application and commercialization in the USA and Europe that utilize either the assimilation capacities of soils or aquatic systems. These systems are designed to contain all effluents and to maximize the influence of natural degrading processes to take advantage of the maximum assimilation capacities of natural processes.

Two natural sludge treatment alternatives are illustrated in Fig. 8-3. Semi-continuous incorporation of liquid sewage sludge into soils that are underdrained to prevent contamination of ground water can assimilate large quantities of sludge slurries. It is likely that the limiting parameter for these systems will be the nitrification rate that can be achieved. This rate appears to be $\approx$110 kg NH_3–N ha^{-1} d^{-1} or greater after the biological system is developed (Jewell, 1982). Any movement of pollutants through a dedicated soil system is captured by the underdrain and returned to the wastewater treatment plant for further processing. A system similar to the above was described by Troemper (1974) and Andrew and Troemper (1975) for two large aerobic wastewater treatment facilities. Although sludge was not applied at the maximum assimilation rates, minimal pollutants were reported in the return flows, and almost all of the nitrogen (N) was destroyed (presumably by high sequential rates of nitrification–denitrification in the soils). This simple and effective sludge minimization treatment was reported to have an operations and maintenance cost of less than $1.00 per ton of sludge.

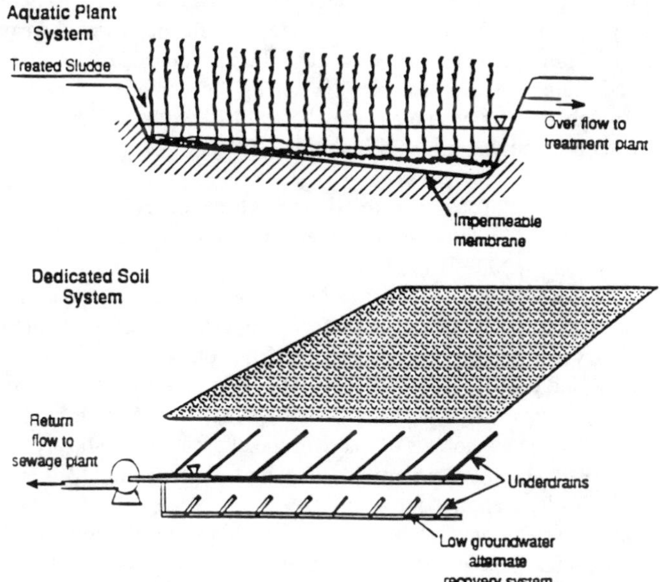

Fig. 8-3. Overview of natural systems that can be used for on-site reduction of sludge slurries in small land areas (<1–3 acres per 10 000 people).

Aquatic plant system sludge assimilation utilizes plants capable of transferring oxygen (O_2) to the root mass. This transfer has been estimated to be >100 kg of O_2 ha^{-1} per d^{-1}. Any excess liquid from this system is also returned to the wastewater treatment plant.

Assimilation capacities for liquid sewage sludge at dry matter of 2 to 8% results in minimum area requirements for the aquatic and soil systems of between one and three acres per million gallons per day of wastewater flow. Hydraulic application rates equivalent to one-tenth of a centimeter per day or ≈37 cm yr^{-1} would be used in typical systems. Although this represents an exceptionally high nutrient loading and a significant organic loading rate, the liquid loading rate is low and evapotranspiration rates are often higher than the application of the slurries.

These systems can be designed to have no effluent discharge for 10 to 20 yr. This long residence time can result in further degradation of refractory organic matter by the biological community. It is known that the humus in top soil is degraded at a low but continuous rate. This would be true also of the refractory organics in the sludge. If the system is harvested every 10 yr, total quantity of sludge generated (of soil-like material) could be as low as 0.2 m^3 per day per 10 000 people. This represents an on-site sludge reduction efficiency of 98% without the use of dewatering technologies beyond sedimentation processes.

Final disposal of the highly stabilized refractory organics in these intensive natural systems must also consider the magnification of components such as heavy metals. Because the stabilized sludge has a high ash content, the resulting final concentration of heavy metals would be lower than one might expect. It is anticipated that most of the solids produced by the system would have a cadmium (Cd) content of <10 mg kg^{-1} dry matter, for example. In addition, biochemical mechanisms may be available to help further reduce toxic organics, such as polychlorinated biphenyls (PCBs).

Pathogen Reduction, Vector Control, and Heavy Metal Reduction

New sludge management regulations provide clear guidance regarding both treatment processes that are acceptable as well as required sludge characteristics for land application (U.S. Environmental Protection Agency, 1993). New and emerging technologies will be judged by a USEPA committee as to their *equivalency* in achieving required treatment. As will be discussed in the next section, the economics of many of the alternatives are costly, and simple and lower cost alternatives are needed to achieve pathogen reduction and metal control prior to land application.

One emerging technology has the potential to be a highly simple system for solids stabilization, pathogen control, and perhaps heavy metal reduction. It is referred to as *liquid composting* or *autothermal aerobic digestion* (ATAD) process (Jewell & Kabrick, 1980; Jewell et al., 1980; U.S. Environmental Protection Agency, 1990). All domestic sewage sludge can be settled to sufficient solid concentrations that will support autoheating to temperatures that destroy pathogens in a simple, aerobic digestion process. This technology was demonstrated at full scale in the late 1970s in the USA (Cummings & Jewell, 1977).

The autoheating of dilute sewage sludge slurries is dependent upon high O_2 transfer efficiency achieved by unique high-shear field, self-aspirating aerators. The technology is fully commercial in Europe, but has yet to be implemented in the USA (U.S. Environmental Protection Agency, 1990). The liquid composting system is easy to construct and operate, and occupies small land areas. As such, it would appear to be more attractive than conventional, high solids aerobic composting. The system is essentially a liquid reactor with adequate aeration horsepower to provide stabilization and autoheating with a foam control application.

An overview of materials flow in an ATAD process for application to a combined biological sludge is illustrated in Fig. 8-4. In this application a significant amount of return flow occurs containing a relatively small amount of solids and $\approx 28\%$ of the total Kjeldahl nitrogen (TKN). This results in an insignificant increase in NH_3–N in the inflow to a sewage treatment plant. Further information would be helpful to compare such variables as dewatering characteristics of these treated slurries with conventional systems.

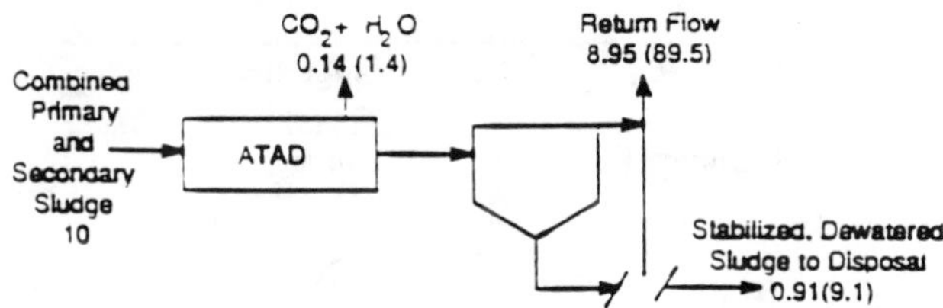

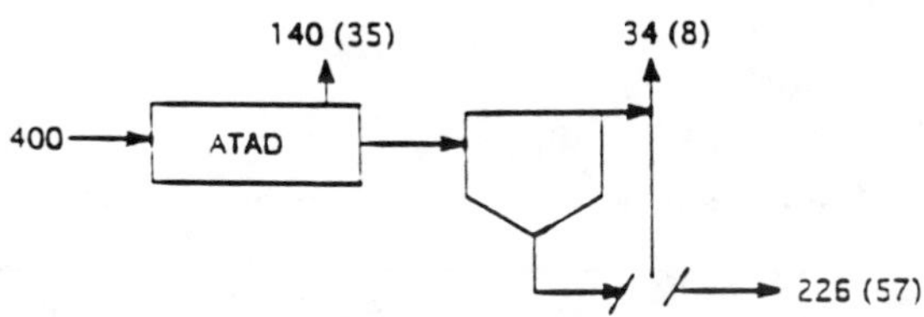

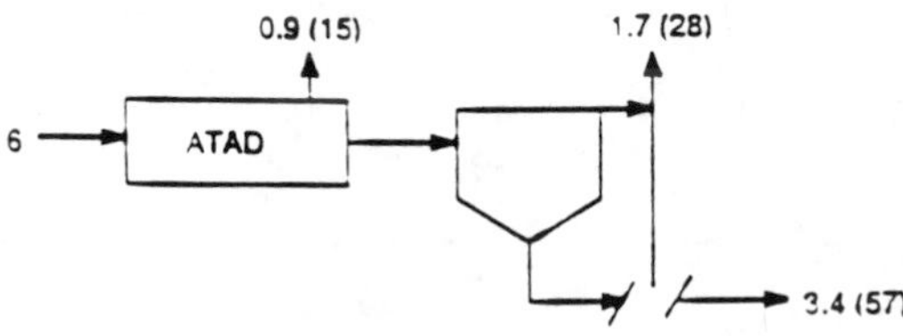

Fig. 8-4. Overview of materials flow in a liquid composting system treating a combined biological wastewater treatment sludge with the following influent characteristics: total wet slurry flow of 10 m^3 per day at 4% total dry matter, 60% of dry matter is organic (volatile solids), and 60% of the organic matter is biodegradable, total Kjeldahl-N is 1.5% of total solids. Values in parentheses are percentage of influent.

The ATAD technology has been determined to be the most cost effective sludge stabilization technology at all sizes with low cost electricity (Wolinski & Bruce, 1984; U.S. Environmental Protection Agency, 1990). Uniform cost comparisons were prepared between the ATAD process, conventional aerobic digestion and anaerobic digestion using actual costs of the European systems (Deeny et al., 1985). In all cases, ATAD was the lowest cost alternative, and in most cases, the capital costs for ATAD were about half the cost of anaerobic digestion.

Table 8-1. Metal concentrations measured in anaerobic digested sludge from three communities prior to detoxification treatment.

	Sludge metal content				
Municipality	Cadmium	Zinc	Copper	Nickel	Lead
	mg kg^{-1} †				
Binghamton	16.1	2,900	749	482	247
Syracuse	27.4	620	871	88	195
Chemung Co.	115.0	2,500	960	43	244

†Dry weight basis.

Table 8-2. Typical metal concentrations in detoxified sludge.

	Sludge metal content				
Municipality	Cadmium	Zinc	Copper	Nickel	Lead
	mg kg^{-1} †				
Binghamton	1.4	200	278	60	82
Syracuse	3.1	141	199	21	46
Chemung Co.	11.9	169	165	10	178

†Dry weight basis.

Table 8-3. Summary of maximum metal removal efficiencies achieved with full-scale testing on anaerobically digested municipal sludges. Values are percentage of initial metal concentrations remaining in the detoxified solids.

	Metal removal efficiency				
Municipality	Cadmium	Zinc	Copper	Nickel	Lead
	%†				
Binghamton	84	92	72	88	72
Syracuse	91	89	72	89	76
Chemung Co.	92	97	86	77	15

†Percentage of initial solubilized.

Although the simplest application of this technology would be in a stand-alone process, it may not be the most effective location for the technology. A logical application would be to use the ATAD alternative following anaerobic digestion. This would enable the cost-effective and efficient anaerobic digestion process to recover about $4,500 annually per 10 000 people of substitute natural gas, while still achieving thermophilic temperatures in the second stage. Full scale tests with anaerobically digested sludge have shown that this series treatment is feasible (Jewell, 1991). Such a series treatment system would also enable existing facilities using anaerobic digestion to retrofit with a pathogen-reducing technology that is simple and easy to operate.

A broad spectrum heavy metal removal technology has also evolved with the development of the ATAD process (Jewell et al., 1980; Jewell, 1991). By controlling the slurry oxidation state, temperature, and pH, the removal of most heavy metals has exceeded 80% with lower efficiencies for copper (Cu) and lead (Pb) (see Tables 8-1, 8-2, and 8-3).

The total chemical requirements for acidification and neutralization of sludge in the heavy metal detoxification process would cost between $20 and $40 per dry ton of treated sludge. This technology applies to sludge stabilized by both aerobic and anaerobic treatment. Although all but 1.5% of domestic sludges meet USEPA ceiling concentration limits for heavy metals, this technology would apply to highly contaminated domestic sludges as well as to industrial sludges.

ECONOMIC CONSIDERATIONS

Just as additional guidance is needed to judge effectiveness of sludge volume and mass reductions on a global basis, costs of sludge management alternatives need further documentation. The best available sludge management cost manuals are nearly a decade old and must be used with caution (U.S. Environmental Protection Agency, 1983, 1985). Good engineering incorporates the best, most reliable, technology for the lowest cost. A uniform and simple cost effectiveness test would be useful to prioritize alternatives as well as to identify areas where additional research and development is needed. Many, if not most, wastewater treatment technologies presently have unacceptably high cost. If cost-effective wastewater treatment is defined as costing less than one percent of the median income of the public served, this would set up the limits on acceptable costs. If the median income was $20,000 per residence per year, total cost of waste treatment would be about $200 per year. This would fix the total cost of wastewater sewage treatment to $0.42 per day per m^3 of treatment capacity ($1.60 per 1000 gal per day).

Upper limits on costs of sludge treatment disposal depend on the distribution of costs to other waste management system components. If 15 to 30% of total costs were allocated to sludge handling, treatment, and disposal, sludge management costs would be between $400 and $800 per metric ton (dry).

Many sludge treatment technologies have total costs greater than the acceptable limits as defined above. Aerobic composting requires capital investment of between $500,000 and $1,000,000 per dry ton per day of processing capacity. Composting costs $645 per dry ton according to costing manuals, and transportation costs can be equal to treatment costs. Thus, the total cost of sludge treatment can easily exceed a simplistic cost effectiveness test such as given above.

Many publications promote the beneficial impact of sludge application to land. The estimated fertilizer value for sludge and the associated dewatering requirements are summarized in Table 8-4. These clearly indicate that the major driving force for land application of sludges will not be its fertilizer value.

If the intensive natural systems can be proven to be reliable and reduce sludge masses by 98% or greater, it is likely that large economic incentives will promote their use. Depending on whether these systems are covered with greenhouses or uncovered, the economics would appear to result in total sludge processing costs varying from less than $100 per metric ton (dry) to $300 per metric ton (dry).

It is difficult to compare costs for septage management to a conventional treatment since the septic tank system is virtually a zero discharge system that discharges drinking water quality to ground water. Sludge management and disposal costs, however, can be estimated. Total costs of pumping and disposing of sludge from a septic tank has typical costs of approximately $70 per tank or per 1000 gallons treated (U.S. Environmental Protection Agency, 1993). This results in a total septage treatment and disposal cost of approximately $500 per metric ton (dry).

Table 8-4. Comparison of nitrogen (N) fertilizer value† of sewage sludge and the fractions of water that must be removed.

Sludge dry matter (wet weight)	N content (per wet ton)	N value (per wet ton)	Water removed (initial present)
%	lb	$	%
2	0.8	0.16	0
5	2.0	0.40	61
10	4.0	0.80	82
15	6.0	1.20	88
20	8.0	1.60	92
25	10.0	2.00	94
35	14.0	2.80	96

†Assumptions include settle sludge produced at 2% total dry matter, N at 2% of the dry matter and it remains constant as water is removed. Value of total N is $0.44 kg^{-1}$ ($0.20 per pound).

If one assumes that the upper limit on annual costs for septic tank systems would be equivalent to $200 (the 1% median salary cost-effectiveness test mentioned earlier), acceptable upper limits of septic tank systems could also be estimated. In this example, the capital cost of the system would be limited to approximately $1,400. In many cases residential septic tank systems can be installed for this amount.

Typical disposal options for septage include discharge to conventional wastewater treatment facilities. In general, this would appear to be a poor policy. Well designed septic tanks achieve good treatment to a tertiary level since they achieve plant nutrient control. Discharge of septage to a publicly owned treatment works (POTW) without nutrient control results in discharge of effluents to surface waters that minimizes the effectiveness of septic tank treatment. It is recommended that alternatives, such as septage farming, be considered as a higher priority than discharge to existing sewage treatment facilities.

DISCUSSION

This chapter provides an overview of several issues that are needed to assist in management and disposal of sewage sludge. Further quantification of sludge generation and its relationship to various processes needs to be carefully understood to prioritize unit process selection, process development, and cost-effectiveness evaluations.

Presently, sewage treatment systems result in the production of sludge that represents nearly two-thirds of the solids generated by the human population. Sludge pumped from septic tanks also represents a large problem. Of the solids deposited in a septic tank, ≈40% require final disposal and proper management.

Since sludge management processing and disposal represents a large part of the operating challenge and cost, it is essential that technologies attempt to minimize and simplify the operation. New technologies appear to be available that would assist in stabilizing and limiting the volumes and masses of sludge generated. The use of the assimilation capacities of aquatic and soil systems would appear to be able to reduce sludge by up to 98% and decrease the residual management problem to once every 10 to 20 yr. This would appear to reduce the sludge management cost by between 50 and 90% of existing cost alternatives.

Other technologies are being developed that simplify the management of sludge while meeting more strict sludge management regulations. One technology that is approaching commercialization is the ATAD process. This is a highly simple technology that can be used to stabilize sludge and destroy pathogens. Further research may confirm that it can be modified to control heavy metals.

A simple cost-effectiveness test is not broadly used nor available for defining available technologies. Although sludge provides valuable fertilizing and moisture management characteristics for soils, it has an insignificant monetary value in relation to high costs of producing it and transporting it to various sites. Thus the value of sludge for land application is not a major driving force in the decision-making process. Sludge management costs often exceed $500 per dry ton. This may exceed a cost-effective definition. It is anticipated that sludge reduction technologies will become available that will reduce management costs.

RESEARCH NEEDS

The following briefly summarize research needs in the area of engineering and cost considerations in sludge management and application to land. Research and development efforts are needed to:

1. Develop a concise quantitative understanding of the magnitude of sludge generation (both wet and dry masses), and the impact of various alternatives on the sludge problem and costs;

2. Define the role of septage management in the overall water quality management picture and encourage (or discourage) use of this zero discharge system and compatible septage management practices. Research and demonstration projects are needed to define superior wastewater treatment characteristics of septic tanks, the impact of discharging septage to municipal treatment systems that have no nutrient removal characteristics, and to promote septage farming;

3. Incorporate anaerobic wastewater treatment as the most effective means of reducing quantities of sludge generated;

4. Incorporate a holistic view of land application of sludges, including compensation for nearby property owners whose activities may be adversely affected by this sludge utilization–disposal option;

5. Define the potential of intensive natural systems (both soil and aquatic systems) to assimilate sludge to document that potential reduction may approach 98%;

6. Define long term degradation rates of sludge refractory organics in heavily loaded soil and aquatic systems;

7. Demonstrate rational aerobic liquid composting system (ATAD) applications that maximize energy production, vector control, and pathogen reduction in a highly simple system — two stage series treatment using mesophilic anaerobic digestion followed by autoheated aerobic digestion;

8. Demonstrate heavy metal removal processes and document the capability to recover and recycle all components of sludge using modification of the liquid composting system;

9. Define heavy metal sludge extraction feasibility using plants;

10. Develop alternative technologies that are capable of reducing sludge generation, such as anaerobic wastewater treatment;

11. Define the role and costs of complete destruction technologies more clearly, e.g., wet oxidation;

12. Define future changes in societal practices that can modify the sludge management problem, i.e., use of large numbers of garbage grinders to avoid increasing solid waste disposal fees, and conversely, use of on-site residential composting operations;

13. Define the potential to use and eliminate sludge in new beneficial ways — slow release fertilizers or feed to vermiculture; and,

14. Update USEPA sludge management and costing manuals to reflect new regulations.

REFERENCES

Andrew, R., and A.P. Troemper. 1975. Characteristics of underflow resulting from cropland irrigated with sewage sludge. Water Pollut. Control Fed. Conf., Miami Beach, FL. 9 Oct. 1975.

Cummings, R.J., and W.J. Jewell. 1977. Thermophilic aerobic digestion of dairy wastes. p. 637–658. *In* R.C. Loehr (ed.) Food, fertilizer, and agricultural residues. Ann Arbor Sci. Publ., Ann Arbor, MI.

Deeny, K., J. Heidman, and J. Smith. 1985. Autothermal aerobic digestion in the Federal Republic of Germany. p. 959–968. *In* Proc. 40th Annual Purdue Univ. Industrial Waste Conf., West Lafayette, IN. Butterworths, Boston.

Jewell, W.J. 1980. Use and treatment of municipal wastewater and sludge in land reclamation and biomass production projects: An engineering assessment. p. 448–480. *In* Utilization of municipal wastewater and sludge - An engineering assessment for land reclamation and biomass production. Symp. Proc. EPA 430/9-81-012. Pennsylvania State Univ. Press, University Park.

Jewell, W.J. 1982. Applying sludge to dedicated land. Biocycle: Sept.–Oct.: p. 42–44.

Jewell, W.J. 1988. Anaerobic sewage treatment. Environ. Sci. Technol. 22:14–21.

Jewell, W.J. 1991. Detoxification of sludges: Autoheated aerobic digestion of raw and anaerobically digested sludges. p. 79–90. *In* H. M. Freeman and P. R. Sferra (ed.) Biological processes. Vol. 3. Innovative hazardous waste treatment technology series. Technomic Publ. Co., Lancaster, PA.

Jewell, W.J., R.J. Cummings, T.D. Nock, E.E. Hicks, and T.E. White. 1993. Energy and biomass recovery from wastewater: Piloting resource recovery wastewater treatment. Gas Research Institute Final Report. GRI–93/0192.1. Gas Res. Inst., Chicago.

Jewell, W.J., and R.M. Kabrick. 1980. Autoheated aerobic thermophilic digestion with air aeration. J. Water Pollut. Control Fed. 52:512–23.

Jewell, W.J., R.M. Kabrick, and J.A. Spada. 1980. Autoheated aerobic thermophilic digestion with air aeration. U.S. Environmental Protection Agency Final Report EPA-600/S2-82-023. U.S. Department of Commerce. NTIS Report PB82-196908.

Troemper, A.P. 1974. The economics of sludge irrigation. p. 115–121. *In* Municipal sludge management. Information Transfer, Rockville, MD.

U. S. Environmental Protection Agency. 1983. Process design manual for land application of municipal sludge. USEPA Rep. EPA-625/I-83-016. U.S. Gov. Print. Office, Washington, DC.

U. S. Environmental Protection Agency. 1985. Handbook estimating sludge management costs. USEPA Rep. EPA-625/6-85010. U.S. Gov. Print. Office, Washington, DC.

U. S. Environmental Protection Agency. 1990. Autothermal thermophilic aerobic digestion of municipal wastewater sludge. Env. Regs. and Tech. USEPA Rep. EPA/625/w-90/007. U.S. Gov. Print. Office, Washington, DC.

U. S. Environmental Protection Agency. 1993. Standards for the use or disposal of sewage sludge. (40 CFR Parts 257, 403 and 503) - Final Rate and Phased-In Submission of Sewage Sludge Permit Application (Revisions to 40 CFR Parts 122, 123 and 501). Final Pub. Distributed by Water Environment Federation. Stock # P0100. U.S. Gov. Print. Office, Washington, DC.

Wolinski, W.K., and A.M. Bruce. 1984. Thermophilic oxidative sludge digestion using air: Performance and costs. Sixth European Sewage and Refuse Symposium, Munich, Germany.

9 Utilization of Risk Assessment in Development of Limits for Land Application of Municipal Sewage Sludge

James A. Ryan

Risk Reduction Engineering Laboratory[1]
U.S. Environmental Protection Agency
Cincinnati, Ohio

Efforts toward rational assessments of risk associated with changes in soils are frustrated by lack of adequate understanding of the dynamics of the interaction of the contaminant and soil, as well as the underlying social ramifications associated with change. The basic concerns of soils, ecology, and toxicology assure that few individuals possess the education and experience to address the complex nature of the contaminant's environmental and toxicological behavior from a technical prospective. Further, the arguments of Simms and Beckett (1987) and Vegter et al. (1988) about the use of science in the resolution of soil quality issues assures that misuse of science in formulation of public policy will occur as it is most certain that some decision will be made, with or without technical input. Decision makers without technical training fail to become informed on the technical issues and must rely on conflicting input from the technical community to formulate the technical issues for input into their decision. Thus, the decision makers are forced to rely on public policy issues to develop their response.

Science may not yield the final answer on the risk issue, but it can and must provide a technical basis "a bell curve with the preponderance of scientific evidence in the middle" on which social issues can be considered to develop the final answer. It must be recognized that the decision to allow anthropogenic changes in soil is a complex issue involving technical as well as policy decisions and that the correct decision cannot be reached if the technical and policy issues are not understood. It becomes apparent that the scientist must communicate

---

[1]Although the research described in this article has been undertaken by the U.S. Environmental Protection Agency, it has not been subjected to Agency review. Therefore, it does not necessarily reflect the views of the Agency.

Sewage Sludge: Land Utilization and the Environment. SSSA Miscellaneous Publication.

what is known along with what is not known and not allow nonscientific factors, e.g., the regulatory outcome, to influence the technical description. Then the informed public through its policy process can determine what is good or bad, how to meet regulatory objectives or serve political purposes. At the same time the technically trained person as well as the policy maker must recognize that change is inevitable and as members of society they have a responsibility to guide change in the way that society dictates. If the scientist merges his good or bad values in his evaluation of the technical issues, the decision is subverted; endlessly confusing rhetoric may result that does not allow the public to have an informed opinion. At the very least the scientist should identify where science and policy have been mixed rather than imply that it is all science.

Mindful of the aforementioned factors, a review of the Clean Water Act 503 methodology to assess the risk from agricultural and nonagricultural land application and distribution and marketing of municipal sewage sludge will be contrasted with the traditional approach of using the distribution of soil background concentration to determine when a soil is polluted. The purpose of this chapter is to present a summary of the risk assessment and policy decisions that led to the U.S. numerical limits for beneficial utilization of sewage sludge on land. For a more complete discussion of the Clean Water Act 503 methodology see U.S. Environmental Protection Agency, 1993a; Chaney and Ryan, 1993; and Ryan and Chaney, 1993.

CLEAN WATER ACT 503 METHODOLOGY

Under authority of Sections 405(d) and (e) of the Clean Water Act as amended (33 U.S.C.A. 1251, et seq.), the U.S. Environmental Protection Agency (USEPA) promulgated a regulation to protect public health and the environment from any reasonably anticipated adverse effects of certain pollutants that may be present in sewage sludge. The Standards for the Use or Disposal of Sewage Sludge (40 CFR Part 503) were published in the Federal Register (U.S. Environmental Protection Agency, 1993b). The regulation establishes requirements for the final use or disposal of sewage sludge in three circumstances. First, the regulation establishes requirements for sewage sludge when sewage sludge is applied to the land for beneficial purposes. Second, the regulation establishes requirements for sewage sludge when sewage sludge is disposed on land by placing it in surface disposal sites. Third, the regulation establishes requirements for sewage sludge when sewage sludge is incinerated. The standards for each end use or disposal practice consist of general requirements, numerical limits on the pollutant concentrations in sewage sludge, management practices, operational standards, frequency of monitoring requirements, recordkeeping requirements, and reporting.

The overall approach utilized for development of sludge pollutant loading limits consisted of the following general components:

- Delineation of pollutants of concern in sewage sludge;

- Identification of potential pathways for exposure and receptors (humans, soil biota, plants, and animals) to these pollutants through land application of sewage sludge;
- Identification of effect endpoints (dose-response relationships) for the receptors and pollutants of concern;
- For each exposure pathway and each pollutant, determine the level of pollutant in the soil which would protect the highly exposed individual–receptor from adverse effects;
- Determine maximum acceptable loading rate of each pollutant based on the most limiting value for all evaluated pathways; and
- From the maximum acceptable loading rates and the sludge concentration distribution of the National Sewage Sludge Survey (NSSS), the pollutant limits (cumulative soil pollutant application limit and maximum allowed sludge pollutant concentration) for the CWA-503 Rule were determined.

Pollutant Evaluated

The USEPA Office of Science and Technology (OST) began to identify pollutants that may pose health or environmental hazards when sewage sludge is used or disposed. To develop this list of pollutants of concern, the following variables were considered: frequency of occurrence, aquatic toxicity, phytotoxicity, human health effects, domestic and wildlife effects, and plant uptake. Originally, four use or disposal practices were identified: land application, landfilling (now called surface disposal), incineration, and ocean dumping. Experts, who were given broad latitude in determining which pollutants to evaluate, met during 1984. They evaluated the potential pollutants of concern for each use or disposal practice by answering the following questions:

- For which pollutants are there sufficient data indicating that such pollutants present a potential hazard if used or disposed by the practice in question?
- For which pollutants are there sufficient data indicating that such pollutants do not present a potential hazard or problem to human health or the environment?
- For which pollutants are there insufficient data to make a conclusive recommendation concerning potential hazard?

Based on the experts' recommendations, 50 pollutants were identified for further analysis. Additionally, the experts designated which environmental exposure pathways were of concern for each pollutant. For land application, 10 environmental pathways and 31 pollutants of concern were identified. From further evaluations an environmental profile, consisted of two sections: a compilation of data on toxicity, occurrence, and fate and effects for the pollutant; and an evaluation of the hazard specific to the environmental pathways for the use or disposal practice of concern was developed for each pollutant (U.S. Environmental Protection Agency, 1985).

In the final regulatory effort 14 organic pollutants and 10 inorganic pollutants were evaluated in the land application Part 503 risk assessment. They include the pollutants identified in USEPA (1985) plus two additional organic pollutants:

Aldrin/dieldrin (total)	Arsenic
Benzene	Cadmium
Benzo(a)pyrene	Chromium
Bis(2-ethylhexyl)phthalate	Copper
Chlordane	Lead
DDT/DDE/DDD (total)	Mercury
Heptachlor	Molybdenum
Hexachlorobenzene	Nickel
Hexachlorobutadiene	Selenium
Lindane	Zinc
N-Nitrosodimethylamine	
Polychlorinated biphenyls	
Toxaphene	
Trichloroethylene	

Subsequent to the completion of the risk assessment, the organic pollutants were deleted from Part 503. The justification for the deletion of those pollutants is presented in Appendix B in the Technical Support Document for Land Application of Sewage Sludge and states, "For an organic pollutant to be deleted from the regulation for a particular use or disposal practice, one of the following three criteria had to be satisfied:

1. The pollutant has been banned for use in the United States; has restricted use in the United States; or is not manufactured for use in the United States.
2. Based on the results of the National Sewage Sludge Survey (NSSS), the pollutant has a low percentage detection in sewage sludge.
3. Based on data from the NSSS, the limit for an organic pollutant in the Part 503 exposure assessment by use or disposal practice is not expected to be exceeded in sewage sludge that is used or disposed."

Table 9-1. Exposure pathways in Part 503 land application risk assessment.

Pathways	Description of the MEI[†]
1. Sludge - Soil - Plant - Human	Consumers in regions heavily affected by landspreading of sludge.
2. Sludge - Soil - Plant - Human	Farmland converted to residential home garden use 5 yr after reaching maximum sludge application.
3. Sludge - Soil - Human	Farmland converted to residential use 5 yr after reaching maximum sludge application with children ingesting soil.
4. Sludge - Soil - Plant - Animal - Human	Households producing a major portion of their dietary consumption of animal products on sludge-amended soil.
5. Sludge - Soil - Animal - Human	Households consuming livestock that ingests soil while grazing.
6. Sludge - Soil - Plant - Animal	Livestock ingesting food or feed crops.
7. Sludge - Soil - Animal	Grazing livestock ingesting soil.
8. Sludge - Soil - Plant	Crops grown on sludge-amended soil.
9. Sludge - Soil - Soil - biota	Soil biota living in sludge-amended soil.
10. Sludge - Soil - Soil - biota - Predator	Animals eating soil biota.
11. Sludge - Soil - Airborne dust - Human	Tractor operator exposed to dust.
12. Sludge - Soil - Surface water-Fish and Human	Water Quality Criteria and human eating fish and drinking water.
13: Sludge - Soil - Air - Human	Humans breathing fumes from volatile pollutants in sludge.
14: Sludge - Soil - Ground water - Human	Humans drinking water from wells.

[†]Most exposed individual.

Exposure Pathways

For the Part 503 land application risk assessment, 14 pathways of exposure were considered. They include the 10 pathways from USEPA (1985) plus four additionally pathways that have since been developed. Not all of the 14 pathways were evaluated for each pollutant. This resulted because preliminary analysis indicated the pathway was not critical for the compound or data needed to make the calculation was not available. For example, only the organic pollutants were evaluated by the vapor pathway (Table 9–1, i.e., Pathway 13).

A separate risk assessment was conducted for certain pathways for managed lands (e.g., agricultural land (includes pasture and range land), lawns, and home gardens) and for unmanaged lands (e.g., forests and reclamation sites). Not every pathway was evaluated for each type of land. For example, the soil ingestion pathway (Pathway 3) was not evaluated for unmanaged land because children are not expected to be exposed for a long period to sewage sludge applied to that type of land.

Health and Environmental Endpoints

Evaluating dose-response data involves quantitatively characterizing the connection between exposure to a chemical (measured in terms of quantity and duration) and the extent of toxic injury or disease. Most dose–response relationships are estimated based on animal studies, because even good epidemiological studies rarely have reliable information on exposure. Therefore, this discussion focuses primarily on dose–response evaluations based on animal data.

There are two general approaches to dose–response evaluation, depending on whether the health effects are based on threshold or nonthreshold characteristics of the chemical. In this context, thresholds refer to exposure levels below which no adverse health effects occur. For effects that involve altering genetic material (including carcinogenicity and mutagenicity), the USEPA's position is that effects occur at very low doses; therefore, they are modeled with a no threshold assumption. For most other biological effects, it is usually (but not always) assumed that *threshold* levels exist.

For nonthreshold effects, the key assumption is that the dose–response curve for such chemicals exhibiting these effects in the human population, achieves zero risk only at zero dose. A mathematical model is used to extrapolate response data from doses in the observed (experimental) range to response estimates in the low–dose ranges. Several mathematical models have been developed to estimate low–dose risks from high–dose experimental risks. Each model is based on general theories of carcinogenesis rather than on data for specific chemicals. The choice of extrapolation model can have a significant impact on the dose–response estimate. For this reason, the USEPA's cancer assessment guidelines recommend the use of the multistage model, which yields estimates of risk that are conservative, representing a plausible upper limit of risk. With this approach, the estimate of risk is not likely to be lower than the true risk (U.S. Environmental Protection Agency, 1986b).

The potency value, referred to by the Carcinogenic Assessment Group as q_1*, is the quantitative expression derived from the linearized multistage model that gives a plausible upper–bound estimate to the slope of the dose–response curve in the low-dose range. The q_1* is expressed in terms of risk–per–dose, and has units of (mg kg^{-1} d^{-1}). USEPA's q_1* values can be found in the Integrated Risk Information System (IRIS), accessible through the National Library of Medicine.

Dose–response relationships are assumed to exhibit threshold effects for systemic toxicants or other compounds exhibiting noncarcinogenic, nonmutagenic health effects. Dose–response evaluations for substances exhibiting threshold responses involve calculating what is known as the reference dose (oral exposure) or reference concentration (inhalation exposure), abbreviated to RfD and RfC, respectively. This measure is used as a threshold level for critical noncancer effects below which a significant risk of adverse effects is not expected. The RfDs and RfCs developed by USEPA can be found in IRIS.

The RfD–RfC methodology uses four experimental levels: no observed effect level (NOEL), no observed adverse effect level (NOAEL), lowest observed effect level (LOEL), or lowest observed adverse effect level (LOAEL). Each level is stated in milligrams per kilogram body weight per day, and all the levels are derived from laboratory animal or human epidemiology data. When the appropriate level is determined, it is then divided by an appropriate uncertainty (safety) factor. The magnitude of safety factors varies according to the nature and quality of the data from which the NOAEL or LOAEL is derived. The safety factors, ranging from 10 to 10 000, are used to extrapolate from acute to chronic effects, interspecies sensitivity, and variation in sensitivity in human populations (Barnes & Dourson, 1988). They are also used to extrapolate from a LOAEL to a NOAEL. Ideally, for all threshold effects, a set of route-specific and effect-specific thresholds should be developed. If information is available for only one route of exposure, this value is used in a route-to-route extrapolation to estimate the appropriate threshold.

The risk reference dose (RfD) utilized as an exposure endpoint for the noncarcinogenic inorganic pollutants in the pathway analysis is the daily intake of a chemical which, during an entire lifetime, is without appreciable risk on the basis of all the known facts. It is apparent that this value is developed to protect the most susceptible members of the population and thus allows greater protection for the majority of the population. Further, it must be recognized that these health endpoints as utilized in this analysis require a chronic lifetime (50–70 yr) exposure for the population at risk.

In the case of environmental endpoints for plants, animals and soil biota, the methodologies are not as well developed, however, an attempt was made to be just as conservative in their selection as was done for humans. The length of exposure required with these was again the lifetime of the receptor (e.g., growing season in the case of annual plants).

Exposure

Exposure evaluation uses data concerning the nature and size of the population exposed to a substance, the route of exposure (i.e., oral, inhalation, or dermal), the extent of exposure (concentration times time), and the circumstances of exposure.

There are two ways of estimating environmental exposure concentrations:

- Directly measuring levels of chemicals (monitoring)
- Using mathematical models to predict concentrations (modeling)

The data that provide the most accurate information about exposure, and the population that is exposed, come from monitoring the population and their environment. Such measurements, however, are expensive and are often limited geographically. The best use of such data is to calibrate mathematical models that can be more widely applied. Estimating concentrations using mathematical models must account not only for physical and chemical properties related to fate and transport, but must also document mathematical properties (e.g., analytical integration vs. statistical approach), spatial properties (e.g., one, two, or three dimensions), and time properties (steady-state vs. nonsteady-state).

Selecting a model for a given situation from the numerous models on fate, transport, and dispersion depends on the following criteria: capability of the model to account for important transport, transformation, and transfer mechanisms; fit of the model to site-specific and substance-specific parameters; data requirements of the model, compared to availability and reliability of off-site information; and the form and content of the model output that allow it to address important questions regarding human or environmental exposures. To the extent possible, selection of the appropriate fate and transport model should follow guidelines specified for particular media where available; for example, the Guidelines on Air Quality Models (U.S. Environmental Protection Agency, 1986a).

In the case of the modeling approach, it is important that a causal link between the population and exposure be established and, ideally, an analysis of population exposure would be necessary. Population analysis involves describing the size and characteristics (e.g., age or sex distribution), location (e.g., workplace), and habits (e.g., food consumption) of potentially exposed human and nonhuman populations.

In defining exposure, the highly exposed individual (HEI) is of critical importance. A HEI is the human, plant or animal that represents a living organism that, because of individual circumstances, has the maximum exposure to a given contaminant for a particular disposal practice. While this concept seems simple, it presents severe methodological problems to a risk assessment. Risk assessment is fundamentally a probabilistic analysis dealing with a random variable. Traditionally, risk assessment has dealt with two extreme ends of the risk scale. One is the low probability–high consequence risk (e.g., nuclear reactor meltdown). The other is the high probability–low consequence risk (e.g., car accidents). The HEI approach that is utilized by USEPA represents another extreme, namely a low probability–low consequence risk. That is, the probability that an HEI as defined actually exists is certainly very small, and it may approach zero. Additionally, the health consequence based on USEPA policy, if this hypothetical person (HEI) does exist, is 10^{-4}, or less for carcinogenic chemicals or no greater than the RfD for noncarcinogenic chemicals (i.e., are designed to protect the sensitive individual from harm). It is possible to discuss the upper 99th percentile (or 90th or 95th), but an improperly defined HEI (the individual with the greatest exposure) is a concept without statistical relevance and represents a bounding estimate whose exposure is irrelevant. When multiple worst case assumptions about the HEI are made, do they lead to the 95th percentile, the 99th percentile, the 99.99999th percentile? At some point, which is a function of the size of the exposed population, there are no individuals left in the exposed (HEI) population. Thus, exposure to this undefinable group is irrelevant as no one is at risk. Therefore, the HEI must be defined and corresponds to a very small, but statistically meaningful, percentage of the population before it is appropriate to create algorithms to attempt to quantify its exposure. Thus, information on the exposed population becomes critical as well as information on chemical concentration and time of contact data (duration of exposure) in completing the objective of exposure assessment (U.S. Environmental Protection Agency, 1991).

In the 503 risk analysis, population size was a function of the pathway and was not easily calculated, however, as the exposure pathways were designed to evaluate the HEI, it would be assumed that they represent a small part of the population and that the exposure to the general population would be much less.

Conversion from Application Rate to Sludge Concentration

For the land application risk assessment the limits were presented in units of mass loading (kg ha^{-1}), except for Pathways 3, 5, and 7, which were concentration based. In order to develop pollutant concentration in sewage sludge (mg kg^{-1}) from the cumulative pollutant loading rate (kg ha^{-1}) it was assumed that the sewage sludge was applied at a rate of 10 Mg ha^{-1} per year for 100 yr (i.e., 1000 Mg ha^{-1} of sewage sludge had been applied). It is important to recognize that the same time frame will be required before the exposure will be as high as estimated.

Pollutant Concentration Limit

The USEPA established the pollutant concentration limit (PCL) (NOAEL sludge) by using the lower of two numbers, the risk assessment derived pollutant sludge concentration or the National Sewage Sludge Survey (NSSS) 99th percentile concentration (i.e., 99% of the sewage sludges are lower in concentration). Further, the USEPA established a ceiling concentration for each pollutant by using the higher of the risk assessment derived pollutant sludge concentration or the (NSSS) 99th percentile concentration. These ceiling concentrations were established to prevent land application of sewage sludges containing high concentration of pollutants, and to prevent *backsliding* of pretreatment programs.

CONCLUSIONS

The 503 methodology utilized is not new or highly sophisticated, but does cause us to rethink our preconceived assumptions, how to best represent data, and what the research priorities need to be. There are many uncertainties in the risk assessment. These include: ignorance (countered with more or appropriate research); data measurements (countered by quality control); data variation (countered by generic analysis of relationships); and data manipulation (countered by evaluation and validation of cogent models). Further, it must be recognized that the approaches and endpoints utilized for each pathway make it impossible to compare safety factors between pathways in more than general terms. Therefore, the outputs must be viewed with healthy skepticism, while at the same time recognized as representing the collective understanding of a technical group who has attempted to remove its good or bad value judgements from the predictions. The risk assessment process used for development of the 503 sewage sludge regulation is without question better than anything done to date and represents the best that could be accomplished within the constraints of time, money and available data .

The alternative use of background soil concentration to establish regulatory levels for soil pollutants is unrelated to environmental risk associated with the level; setting soil pollutants based on background soil concentration must be viewed as arbitrary and unnecessarily conservative. In comparison, the CWA-503 methodology allows for a much broader technical understanding of the environmental risk. Additionally it allows an understanding of the scientific data and the areas that would most benefit from further research. Therefore, the CWA-503 methodology has provided the regulatory process with the technical evaluation of risk to include with the policy issues in reaching the regulatory conclusion.

REFERENCES

Barnes, D.G., and M. Dourson. 1988. Reference dose (RfD): Description and use in health risk assessments. Regul. Toxicol. Pharmacol. 8:471–486.

Chaney, R.L., and J.A. Ryan. 1993. Heavy metals and toxic organic pollutants in MSW-compost: Research results on phytoavailability, bioavailability, fate, etc. p. 451–506. *In* H.A.J. Hoitink and H.M. Keener (ed.) Science and engineering of composting: Design, environmental, microbiological and utilization aspects. Renaissance Publ., Worthington, OH.

Ryan, J.A., and R.L. Chaney. 1993. Regulation of municipal sludge under clean water act section 503: A model for exposure and risk assessment for MSW-Compost. p. 422–450. *In* H.A.J. Hoitink and H.M. Keener (ed.) Science and engineering of composting: Design, environmental, microbiological and utilization aspects. Renaissance Publ., Worthington, OH.

Simms, D.L., and M.J. Beckett. 1987. Contaminated land: Setting trigger concentrations. Sci. Total Environ. 65:121–134.

U.S. Environmental Protection Agency. 1985. Summary of environmental profiles and hazard indices for constituents of municipal sludge. USEPA, Office of Water Regulations and Standards, Wastewater Criteria Branch, Washington, DC.

U.S. Environmental Protection Agency. 1986a. Guidelines on air quality models (revised). EPA/OAQPS-450/2-87-027R. USEPA, Washington, DC.

U.S. Environmental Protection Agency, 1986b. Guidelines for carcinogen risk assessment. Part II. U.S. Environmental Protection Agency. Federal Register 51(185):33992–34003. U.S. Govt. Print. Office, Washington, DC.

U.S. Environmental Protection Agency. 1988. Reference Dose (RfD): Description and use in health risk assessments. Integrated Risk Information Systems (IRIS). Online. Intra Agency Reference Dose (RfD)

Work Group, Office of Health and Environmental Assessment, Environmental Criteria and Assessment Office, Cincinnati, OH.

U.S. Environmental Protection Agency. 1991. Guidelines for exposure assessment. Draft final. Risk Assessment Forum. USEPA, Washington, DC.

U.S. Environmental Protection Agency. 1993a. Technical support document for land application of sewage sludge. Vol.I. PB93-110575, Vol.II PB93-110583. NTIS, Springfield, VA.

U.S. Environmental Protection Agency. 1993b. Standards for the use or disposal of sewage sludge. Federal Register 58(32):9248–9415. U.S. Govt. Print. Office, Washington, DC.

Vegter, J.J., J.M. Roels, and H.F. Bavinck. 1988. Soil quality standards: Science or science fiction. p. 309-316. *In* K. Wolf et al. (ed.) Contaminated soils. Kluwer Academic Publ., Boston.

10 Production, Use, and Creative Design of Sewage Sludge Biosolids

John M. Walker

Office of Wastewater Enforcement and Compliance
U.S. Environmental Protection Agency
Washington, DC

Research on and experience concerning the benefits and requirements for safely using municipal sewage sludge biosolids are perhaps more extensive than for any other municipal waste material. This accumulated knowledge has lead to the establishment of risk-based rules for biosolids use, (40 CFR Part 503, U.S. Environmental Protection Agency, 1993), the U.S. Environmental Protection Agency's (USEPA) policy promoting the beneficial use of sludge biosolids, and operating projects that have safely and beneficially used biosolids for many years. Today in some states (e.g., Maryland) ≈90% of the biosolids generated within their jurisdictions are beneficially used. Increasingly, other states have agreed to allow use of Class B biosolids shipped long distances from out of state (e.g., anaerobically digested biosolids shipped by truck from the city of Los Angeles to Arizona and by rail from New York City to Colorado and Texas).

This enhanced knowledge about the use of biosolids makes the important goal that soils be maintained and improved to allow sustained agricultural production of food and feed crops more achievable. It also provides a basis for the development of designer waste materials for specific enhanced agronomic and environmental benefit.

This chapter discusses the useability of a variety of different sludge biosolid products that are available today. The chapter addresses features of each of these products that impact the way they are most often used. Finally, the chapter will briefly discuss needed research and a proposed national consortium. The national consortium would conduct and give guidance on research and help facilitate the sharing of knowledge to further sustained agronomic use of urban and rural wastes.

Sewage Sludge: Land Utilization and the Environment. SSSA Miscellaneous Publication.

BIOSOLID PRODUCTS

A series of different biosolid products are listed in Tables 10-1 and 10-2. In general, the processing of biosolid products necessarily influences what the end product will look like; how publicly acceptable the biosolid product will be; its potential for malodor; its pathogen classification; its content of nitrogen (N), phosphorus (P), potassium (K), lime, and metals; when and how it can be stored and used; and restrictions on its end use.

Pathogen and Pollutant Limitations

The Part 503 rule provides that biosolid products with low pollutant content (less than or equal to in mg kg^{-1}: arsenic (As) = 41, cadmium (Cd) = 39, chromium (Cr) = 1200, copper (Cu) = 1500, lead (Pb) = 300, mercury (Hg) = 17, molybdenum (Mo) = 18, nickel (Ni) = 420, selenium (Se) = 36, and zinc (Zn) = 2800), Class A pathogen reduction (virtual absence of pathogens), and vector attractiveness reduced are the easiest to handle and use. In fact, such low pollutant, Class A, stabilized products are considered a product to be used just like any other organic nutrient and tilth enhancer and are not subjected to further regulation under the Part 503 rules. This is because the risk from the low amounts of pollutants added with these sewage sludge biosolids is so minimal that their use does not have to be tracked. The USEPA has referred to such biosolid products in guidance as *exceptional quality* (EQ) biosolids. Most EQ biosolids produced to date have been treated by composting, heat-drying, or alkaline stabilization.

If low-pollutant biosolids have only Class B pathogen reduction (reduced, but not pathogen-free content), some additional Part 503 restrictions on site access, grazing, and crop harvesting are required to assure safe use. Just as for EQ biosolids, however, the cumulative amounts of pollutants applied to lands in these low pollutant Class B biosolids do not have to be tracked.

Biosolids, containing any pollutants at levels in excess of those listed previously (and taken from Table 3 in the Part 503 rule), can still be safely land applied. The Part 503 rule contains the additional requirement that the cumulative amount of pollutants added to a site not exceed maximum levels (listed in Table 2 of the Part 503 rule). Before the new Part 503 sewage sludge regulation, except for bagged and containerized products, the use of all biosolids on land (irrespective of the levels of contained pollutants) was on a cumulative pollutant loading basis. The new part 503 rule recognizes the safety of low-pollutant biosolids, eases off on their regulation, and encourages their production and use.

Table 10-1. Useability characteristics of selected biosolid products.

Biosolid	How	Characteristics		USEPA Part 503 rule	
product[†]	stabilized	Nitrogen	Physical	Pathogen class	Management practice limitations
		%			
Liquid	Aerobic digestion Anaerobic digestion Lime stabilized	3 to 9 3 to 7 2 to 4	Inject or surface apply as liquid	B	Must meet management practices; and site, crop, and grazing restrictions plus may have malodor
Dewatered	Aerobic digestion Anaerobic digestion Lime stabilized	2 to 5 2 to 5 1 to 3	Rewet-inject or surface apply as cake	B	Must meet management practices; and site, crop, and grazing restrictions plus may have malodor
Air-dried	Aerobic digestion Anaerobic digestion Lime stabilized	2 to 4 2 to 4 1 to 3	Easy to handle and spread	B	Must meet management practices; and site, crop, and grazing restrictions
Alkaline-treated	Process with kiln dust Without kiln dust	1 to 2 1 to 2	Consistency of wet ag lime	A/B	None, assuming low pollutants; if B, must meet management practices; and site, crop, and grazing restrictions plus may have malodor
Composted	Many variations of agitated and static aerated processes in and out of vessels	1 to 2	Easy to store and handle; peat-like	A	None, assuming low pollutants; malodor and bioaerosol concerns during composting
Heat-dried	Dried by indirect, evaporative, and direct heating processes	2 to 10	Sold in pelletized form	A	None, assuming low pollutants; malodor if raw-dried product or if product gets wet when stored

[†]If a low-pollutant product (Table 3 in Part 503 rule) then cumulative metals do not have to be tracked.

Table 10-2. End uses of the biosolid products.

Biosolid product[†]	End use	Advantages	Disadvantages
Liquid	Liquid fertilizer on pasture and row crops, forests, reclamation, etc.	Least costly to produce; higher N content; can store in lagoons; anaerobic processing yields energy	Higher cost to transport long distances; malodor potential; management practices plus site restrictions; aerobic processing uses energy
Dewatered	Fertilizer on pasture and row crops, forests, for reclamation, etc.	Less cost to transport; apply as cake or rewet to inject; anaerobic processing yields energy	More cost to dewater; malodor potential; management practices plus site restrictions; aerobic processing uses energy
Air-dried	Fertilizer on pasture and row crops, forests, for reclamation	Less cost to dewater than mechanical; easy to store and get uniform application	Requires more space and time to dewater and is climate dependent; management practices plus site restrictions
Alkaline-treated	Used as ag lime and artificial soil for daily landfill cover	Quick to get an operation going; relatively inexpensive when not control odor; provides nutrients + lime.	High lime not desirable for some uses; potential malodor when processed and stored
Composted	Valuable as soil conditioner; many horticultural, landscape and nursery uses	High public acceptance of product; peat substitute with slow release of nutrients; easy to handle and store	Process concerns are odor and bioaerosols; relatively expensive to produce
Heat-dried	Mostly used at low rates as nutrient source, carrier in chemical fertilizer	Product contains high N; easy to store and use	Must be kept dry when stored; strong odor if heat-dried primary product; relatively expensive to produce

[†]If a low-pollutant product (Table 3 in Part 503 rule) then cumulative metals do not have to be tracked.

Costs and Trade Offs

The cost of producing biosolid products is generally greater with increased treatment and drying. There are of course many trade-offs, some of which are indicated in Table 10-2. For example:

1. Hauling (except for short hauls) is more costly for liquid than dewatered and heat-dried biosolid products,

2. Facilities for anaerobicly digesting biosolids require more capital investment and construction time than for aerobic digestion or especially for alkaline stabilization without odor control (e.g., there have been situations where portable dewatering and alkaline stabilization units have been set up for extended operation within 1 or 2 wk, compared with 1 or 2 yr for constructing and bringing into operation other forms of treatment), and

3. Storage is much easier for composted and heat-dried than for other forms of biosolids.

Product Uses

The various biosolid products are useful for different end-uses. Some examples of uses for drier products include:

1. Composted biosolids are often used to establish and maintain turf, in nurseries, and as horticultural potting media and have great public acceptance,

2. Alkaline-stabilized biosolids are used for agriculture and disturbed-land reclamation or as artificial soil for daily cover in landfills, and

3. Heat-dried biosolids are costly to produce, but have the highest fertilizer value of the dried products. Heat-dried biosolids contain the least amount of water and are often used as filler in chemical fertilizers or as a fertilizer for lawns and citrus crops.

Product Concerns

It is abundantly clear that outstanding progress has been made in understanding and enhancing the ability to beneficially use biosolids. For example, just consider the 20 highly-documented years of excellent investigation in the biosolids-treated Rosemount Watershed. Consider also the Part 503 rule that was issued after extensive revision in data and risk assessment. There is no doubt in my mind that the Part 503 rule is the most comprehensive and best rule ever developed to govern biosolids use. The knowledge gained about risk assessment during the development of the Part 503 rule has taken the risk assessment process to a totally new higher level.

Can we say, therefore, that little more biosolids research is needed? The answer is clearly no. One reason is the need to address remaining concerns. Public concern has been heightened by the issuance of the part 503 rule. Regulatory concern has been heightened by the inter- and intra-agency review of the rule prior to issuance and the need for state acceptance of the rule with or without modification. Perceived concerns include the belief by some that biosolids use will result in the death of farmers and their livestock, may result in the transmittal of the HIV virus, will dangerously increase the level of Pb in the environment, and would necessitate the need for a 2 to 5 mile barrier around a biosolid composting facilities due to bioaerosols.

A second reason is to help resolve debate about other issues related to biosolids use. For example, is the acceptable limiting level for Cd in a *clean* biosolids product 39, 25, or 21 mg kg^{-1}? How determinant should the European Economic Community policy-derived Cd limit in grain be in setting USA biosolid use standards? Should the Mo limit for *clean* biosolids be more or less than the current 17 mg kg^{-1}? Did the recent risk assessment adequately considered acid soils data, dairy farming, and ecosystems when the Part 503 rules were established?

THE FUTURE

Additional research would help achieve the enormous potential economic and social benefits from safe use of urban and rural wastes, agriculture will benefit from superior improved products, urban dwellers will benefit from safer bagged products and their pollutant-binding capabilities, and jobs will be provided by a healthy new waste processing industry. Without research and expertise, however, important opportunities for use will not be developed and rules governing use of wastes will be overly stringent. The pool of experts with knowledge of advances in risk assessment as a basis for such rules is very limited. The experts with this knowledge helped USEPA avoid a *no change in the environment*, below-detectible-limit policy and *instead* follow a reasonable-risk science-based approach to determine an acceptable level of pollutants in soil following biosolids addition.

The entrance of new scientists into the subject area has continued to dwindle. There has been little-to-no focused Federal presence or money for research in this area for nearly a decade. A national consortium is needed that will provide training, research and assistance to those seeking to make sustained agronomic use of waste products. Oversight of such a national consortium by industry, municipal, regulatory, environmental, citizen, researcher, banker, and farmer representatives is essential in providing the necessary direction and resources needed for research and technical assistance.

RESEARCH AND TECHNICAL ASSISTANCE NEEDS

1. Help design and test the useability of tailor-made waste products. For example:

- Demonstrate and show how Pb problems in urban soils may be partially solved by use of biosolids on high Pb soils to reduce the bioavailability of the soil Pb;
- Explain how to design products to curtail soil and plant diseases;
- Combine wastes with different desirable properties for a given end-use; and
- Learn more about how organic amendments in soil impact the rhizosphere and enhance the uptake of conventionally applied soil nutrients, the quality of food and feeds, and resistance of plants to disease.

2. Evaluate and minimize bioaerosols and odors from composting facilities that has threatened their siting.

3. Provide the scientific understanding to address liability concerns of lending institutions that are threatening the cutoff of money via loans to farmers who make use of wastes on their land. These liability fears are intensified because in some instances *clean* wastes, such as the low-pollutant biosolids, have greater permitted pollutant contents than hazardous waste clean-up sites. This results in part from a lack of understanding about differences in bioavailability.

4. Help transmit knowledge to overcome the opposition of scientists to biosolids use on agricultural land in states with the least experience in such use.

5. Provide the basis for assessing the release of available nutrients (e.g., N) and pollutants (e.g., metals) from different biosolid products.

6. Help establish and apply uniform nutrient management requirements for all soil chemical and organic fertilizing and amending products.

7. Help fuse or defuse the supposition that biosolids use on soils is tantamount to placement of a time bomb that will release tightly bound metals as soon as the soil becomes acid after little or no management, especially in sensitive unmanaged ecosystems.

8. Improve the ability to evaluate ecological as well as agronomic risks from waste product use on land. Probabilistic analyses can assist in determining more realistic standards.

The biosolids-treated University of Minnesota's Rosemount Watershed (see Chapters 17 through 21) is an ideal location to study long-term impacts of biosolids use, e.g., the extent of change in bioavailability of biosolids-added metals over time, both with and without subsequent management of the ecosystem.

REFERENCES

U.S. Environmental Protection Agency. 1993. Standards for the use or disposal of sewage sludge. Federal Register 58(32): 9248-9415. U.S. Gov. Print. Office, Washington, DC.

11 Ecological Aspects of Land Spreading Sewage Sludge

Anne Fairbrother

Environmental Research Laboratory
U.S. Environmental Protection Agency
Corvallis, Oregon[1]

C. M. Knapp

Technical Resources
Davis, California[2]

Forests, rangelands, and agricultural areas generally are managed for commodity production including timber, livestock, and food crops, respectively. Coincident with the managed species there exists a complex, interacting web of naturally occurring plants, animals, and soil biota. Some of these are considered pests as they out compete or predate upon the desired species. Many other species of native plants and animals, however, are valued for their aesthetic or spiritual qualities, for game or subsistence hunting, or for their contributions to the functioning of the larger ecosystem. Plants help to prevent soil erosion and to fix carbon dioxide (CO_2), thereby contributing to the production of clean water and air. Animals distribute seeds and maintain the balance of the plant community through foraging pressures. Soil organisms are essential for the maintenance of soil tilth. Therefore, when the Clean Water Act was amended in 1977 to direct the U.S. Environmental Protection Agency (USEPA) to develop regulations containing standards for the use or disposal of sewage sludge, language was included to direct the Agency to identify pollutants that may be injurious to environmental endpoints as well as human health.

Native plants and wildlife are at risk due to direct toxicity of contaminants in sludge and through indirect effects resulting from changes in the plant and animal communities. Vegetative community development patterns are disrupted by the addition of sewage sludge. The most marked effect is a change in the apparent successional stage to an earlier state, characterized by increased annual plant species over perennial species and increased net primary

[1]Current address: **e**cological **p**lanning and **t**oxicology, inc., 5010 SW Hout St., Corvallis, OR 97333.
[2]Current address: 221 Humboldt Ave., Davis, CA 95616

Sewage Sludge: Land Utilization and the Environment. SSSA Miscellaneous Publication.

productivity (Hyder & Barrett, 1985; Buchgraber & Schechtner, 1989; Bollinger et al., 1991). Increased biomass (Dowdy & Ham, 1977; Du Pont de Nemours and Co., 1985; McLeod et al., 1986; Moffet & Matthews, 1991) and lower species diversity, richness and evenness values (Hyder & Barrett, 1985; Bollinger et al., 1991) also occur on land application sites. Forage grasses increased in proportion to legumes and other plants in a sludge-treated temperate grassland (Buchgraber & Schechtner, 1989). Herbaceous vegetation showed greater response to sludge addition than did shrub-vine and woody vegetation components of a loblolly pine plantation treated with sludge (McLeod et al., 1986). Sludge-treated arid grasslands showed decreased plant density, species richness, and species diversity (Fresquez et al, 1990).

Vegetative growth rates of numerous plant communities were increased by sludge applications, including grasses and legumes (Seaker & Sopper, 1988; Heinicke, 1989), desert shrubs (Sabey et al., 1990) and both overstory trees and understory shrubs in loblolly pine forests (McLeod et al., 1986). In addition, grass and legume detritus production and organic matter increased after sludge application (Seaker & Sopper, 1988; Heinicke, 1989). Virginia pine produced more biomass following sludge treatment at restoration sites than did either loblolly or shortleaf pine (McNab & Berry, 1985).

Additional vegetation responses to sludge amendment include reduction in browse quality for deer of overstory trees and understory shrubs in loblolly pine forests (McLeod et al., 1986) and reduction in drought-tolerance, although the sludge-amended communities responded more quickly to alleviation of drought than did the non-amended communities (Bollinger et al., 1991).

Direct toxic effects of sewage sludge on wildlife are unlikely in the herbivorous or omnivorous food chains, but may be a problem in the detritivore food chain. Bioaccumulation of metals has been shown to reach potentially toxic concentrations in the tissues of woodcock (*Philohela minor*) (Woodyard & Haufler, 1991) and moles (*Talpa europea*) (Ma, 1987), both of which represent relatively long-lived vertebrate predators in detritivore food chains. Cadmium (Cd) was of particular concern as it accumulated to potentially lethal levels in renal and hepatic tissues of woodcock fed earthworms grown on sludge-amended soils. Similarly, elevated Cd concentrations were found in moles captured on sludge-treated sites (Ma, 1987), and lead (Pb) and zinc (Zn) were also occasionally elevated. Evaluation of the insectivorous food chains have shown that, although liver and kidney concentrations of Cd in shrews (*Sorex trowbridgii*) and shrew-moles (*Neurotrichus gibbsii*) from sludge-amended sites were significantly elevated, no lesions or other signs of heavy metal toxicity were evident (Hegstrom & West, 1989).

Indirect effects of sewage sludge applications on wildlife species do occur, however, through alteration of habitat conditions (Dominguez, 1987), quality and quantity of food resources, and changes in vegetative composition and cover (Haufler & West, 1986). In addition to increased competitiveness of annual over perennial plant species (see above), changes in the nutritional quality of food grown on sludge-amended soils has been documented (Haufler & West, 1986), including increased protein, nitrogen (N), and phosphorus (P) content of food. Alberici et al. (1989) documented dietary preference shifts in meadow voles (*Microtus pennsylvanicus*) resulting from differential metals contamination of forage. Increased forage amount and quality has been shown to increase wildlife species diversity and richness on sludge-amended forest sites in Michigan and to improve winter food availability for Columbian black-tailed deer (*Odocoileus hemionus*) in Washington (Haufler & West, 1986; Fresquez et al., 1990).

Numerous studies have indicated that sewage sludge application results in changes in the composition, activity, and biomass levels of soil microflora and microfauna, although the specific response depends on soil-type (Coppola, 1986; Sanders & Adams, 1987; Coppola et al., 1988; El-Husseiny et al., 1988) and conditions of sludge application (Balzer & Ahrens, 1990; Sutton et al, 1991). For example, volcanic soils are more susceptible to reduced ammonification caused by Cd-contaminated sludge than are iron (Fe)-rich soils (Coppola, 1986); fungal population sizes increased in calcareous (clay loam), but not in alluvial (clay) soils (El-Husseiny et al., 1988). More importantly, the N-fixing capacity of mycorrhizae diminished in sludge-amended soil (Heckman et al., 1986; Coppola et al., 1988; Koomen et al., 1990). Soils with metals from applied sewage sludge grew plants with lower concentrations of N, smaller yields, and less well developed and ineffective root nodules than noncontaminated soils, even though the metal concentration of the foliage was low and no phytotoxicity was observed (McGrath et al., 1988). Chaudri et al. (1992) showed that metal tolerant mycorrhizae did not fix N and that N-mycorrhizae were not metal tolerant.

The USEPA is proposing to conduct a formal ecological risk assessment for sewage sludge constituents (including nutrients), integrating environmental exposure profiles with information about ecological effects (abstracted above). If funding is secured, forest, rangeland, and agroecosystem models would be developed for which ecological risks could be estimated and described in terms of selected assessment endpoints. Data gaps would be identified and uncertainties quantified, where possible. A limited amount of field work would be conducted to verify the ecosystem interactions included in the model and to fill large data gaps. Data visualization tools could be used to explain *if-then* scenarios in order to predict the effects of changing the rate of sewage sludge application on the ecological endpoints of concern. Risk managers could use this information to establish allowable loading rates for pollutants.

REFERENCES

Alberici, T.M., W.E. Sopper, G.L. Storm, and R.H. Yahner. 1989. Trace metals in soil, vegetation, and voles from mine land treated with sewage sludge. J. Environ. Qual. 18:115–120.

Balzer, W., and E. Ahrens. 1990. The effect of long-term sewage sludge application on the microbial activity in a silty loam. Reihe Kongressber. Verb. Dtsch. Landwirtsch. Unters. Forschungsans. 30:479–484.

Bollinger, E.K., S.J. Harper, and G.W. Barrett. 1991. Effects of seasonal drought on old-field plant communities. Am. Midl. Nat. 125:114–125.

Buchgraber, K., and G. Schechtner. 1989. Effectiveness of sewage sludge on grassland, especially on soil and plants. Veroff. Bundesans. Alpenlandische Landwirtsch. Gumpenstein. 11:35.

Chaudri, A.M., S.P. McGrath, and K.W. Giller. 1992. Metal tolerance of isolates of *Rhizobium leguminosarum* biovar *trifolii* from soil contaminated by past applications of sewage sludge. Soil Biol. Biochem. 24:83–88.

Coppola, S. 1986. Summary of investigations in Italy into effects of sewage sludge on soil microorganisms. p. 72–79. *In* R.D. Davis et al. (ed.) Factors influencing sludge utilization practices in Europe. Elsevier, Amsterdam.

Coppola, S., S. Dumontet, M. Pontonio, G. Basile, and P. Marino. 1988. Effect of cadmium-bearing sewage sludge on crop plants and microorganisms in two different soils. Agric. Ecosyst. Environ. 20:181–194.

Dominguez, S. 1987. Effects of land-applied municipal sewage sludge on wildlife. USEPA Environ. Res. Lab., Corvallis, OR.

Dowdy, R.H., and G.E. Ham. 1977. Soybean growth and elemental content as influenced by soil amendments of sewage sludge and heavy metals: Seedling studies. Agron. J. 69:300–303.

Du Pont de Nemours and Co. 1985. Sludge application program at the Savannah River plant. Du Pont de Nemours (E.I.) and Co., Aiken, SC. Savannah River Lab.; U.S. Department of Energy, Washington, DC. Forest land applications symposium, Seattle, WA. 25 June 1985.

El-Husseiny, T.M., M.K. Sadik, and N.M. Badran. 1988. Interactions between soil C/N ratio and saprophytic microorganisms in alluvial and calcareous soils. Egypt. J. Microbiol. 23:109–122.

Fresquez, P.R., R.E. Francis, and G.L. Dennis. 1990. Soil and vegetation responses to sewage sludge on a degraded semiarid broom snakeweed/ blue grama plant community. J. Range Manage. 43:325–331.

Haufler, J.B., and S.D. West. 1986. Wildlife responses to forest application of sewage sludge. p. 110–116. *In* D.W. Cole et al. (ed.) The forest alternative for treatment and utilization of municipal and industrial wastes. Univ. of Washington Press, Seattle.

Heckman, J.R., J.S. Angle, and R.L. Chaney. 1986. Soybean nodulation and nitrogen-fixation on soil previously amended with sewage-sludge. Biol. Fertil. Soils 2:181–185.

Hegstrom, L.J., and S.D. West. 1989. Heavy-metal accumulation in small mammals following sewage-sludge application to forests. J. Environ. Qual. 18:345–349.

Heinicke, D. 1989. Spread of nematodes with sludge and sewage. Kartoffelbau. 40:221–224.

Hyder, M.B., and G.W. Barrrett. 1985. The effects of sewage-sludge and fertilizer on plant-species diversity and primary productivity in a 6th-year old-field community. Ohio J. Sci. 85:89–89.

Koomen, I., S.P. McGrath, and K.E. Giller. 1990. Mycorrhizal infection of clover is delayed in soils contaminated with heavy metals from past sewage sludge applications. Soil Biol. Biochem. 22:871–873.

Ma, W. 1987. Heavy metal accumulation in the mole, *Talpa europea*, and earthworms as an indicator of metal bioavailability in terrestrial environments. Bull. Environ. Contam. Toxicol. 39:933–938.

McGrath, S.P., P.C. Brookes, and K.E. Giller. 1988. Effects of potentially toxic metals in soil derived from past applications of sewage sludge on nitrogen fixation by *Trifolium repens* L. Soil Biol. Biochem. 20:415–424.

McLeod, K.W., C.E. Davis, K.C. Sherrod, and C.G. Wells. 1986. Understory response to sewage sludge fertilization of loblolly pine plantations. p. 308–323. *In* D.W. Cole et al. (ed.) The forest alternative for treatment and utilization of municipal and industrial wastes. Univ. of Washington Press, Seattle.

McNab, W.H., and C.R. Berry. 1985. Distribution of aboveground biomass in three pine species planted on a devastated site amended with sewage sludge or inorganic fertilizer. For. Sci. 31:373–382.

Moffet, A.J., and R.W. Matthews. 1991. The effects of sewage sludge on growth and foliar and soil chemistry in pole-stage Corsican pine at Ringwood Forest, Dorset, UK. Can. J. For. Res. 21:902–909.

Sabey, B.R., R.L. Pendleton, and B.L. Webb. 1990. Effect of municipal sewage sludge application on growth of two reclamation shrub species in copper mine soils. J. Environ. Qual. 19:580–586.

Sanders, J.R., and T.M. Adams. 1987. The effects of pH and soil type on concentrations of zinc, copper, and nickel extracted by calcium chloride from sewage sludge-treated soils. Environ. Pollut. Ser. A. 43:219–228.

Seaker, E.M., and W.E. Sopper. 1988. Municipal sludge for minespoil reclamation. II. Effects on organic matter. J. Environ. Qual. 17:598–602.

Sutton, S.D., G.W. Barrett, and D.H. Taylor. 1991. Microbial metabolic activities in soils of old-field communities following eleven years of nutrient enrichment. Environ. Pollut. Ser. A. 73:1–10.

Woodyard, D.K., and J.B. Haufler. 1991. Risk evaluation for sludge-borne elements to wildlife food chains. *In* The environment, problems, and solutions. Garland Publ., New York.

12 Gaining Public Acceptance for Biosolids

Jane B. Forste

Bio Gro Systems
Annapolis, Maryland

Those of us involved in public acceptance efforts recognize that scientific data and information about risk assessment do not necessarily result in public acceptance for the projects we develop. We are told that: (i) perception equals reality (i.e., a perceived risk is just as real to the person concerned about it as is scientifically derived risk assessment) and (ii) in decision-making, perception outweighs reality. We need to remember, however, that perceptions about many subjects can and do change with time and with accurate, credible communication efforts.

The beneficial use of biosolids provides a case in point regarding risk perception. A 20-yr compilation of extensive, detailed, scientific data has clearly established that the treated solids from municipal wastewater can safely be used on land in a variety of ways. The quality standards for such safe beneficial use are based on protective assumptions about the impact of biosolids on the environment, crops, animals, and humans. Why then does a negative perception about the use of this material persist? Part of the answer lies in lack of understanding, the result of a very common and natural human apprehension. The average citizen knows little or nothing about wastewater treatment, the pretreatment programs that protect the integrity of this process, and the biological and chemical composition of the treated solids.

To develop a dialogue with communities where biosolids will be processed or used, it is important to establish the link between our national commitment to clean water and the necessity to manage the solids from wastewater treatment. The environmental issues that relate to municipal wastewater solids and treatment are very different from those involving hazardous wastes; these issues must be clarified to win public confidence and support for beneficial use projects.

Sewage Sludge: Land Utilization and the Environment. SSSA Miscellaneous Publication.

BUILDING ALLIANCES

Many public acceptance campaigns tend to focus efforts on a somewhat vague effort to convince the *general public*. While public acceptance must include general information and outreach, a positive beneficial use program begins with *local* acceptance. Without this component, the risk of negative public opinion is virtually overwhelming.

Local acceptance must be achieved for both facilities (e.g., composting) and operational projects (e.g., land application). The approaches may differ, but the underlying methods of open communication and responsiveness are more likely to result in alliances and support.

For land application programs, the farming community is the obvious and first source of allies in the effort to gain and keep public acceptance. Agricultural organizations, extension agents, and individual farmers can all help to provide the strong local support that makes programs work. Their participation can also be a major component in developing the approval and understanding of local officials, who may or may not be members of the farming community. Since land application provides a significant resource to agriculture, that value must be recognized and articulated by the farmers who use biosolids.

The agricultural value of biosolids can only be fully realized through appropriate field management that involves coordination with individual farming practices. The timing, method(s) of application, and careful attention to housekeeping details during operations will help to insure continued cooperation. By being responsive first and foremost to the farmers' needs, the underlying support for a land application program will be maintained.

It is important to recruit and train highly motivated technical staff with backgrounds in such disciplines as agronomy, soil science, natural resources, and environmental science. These specialists contact and work one-on-one with farmers. They obtain soils information, cropping history, regulatory mapping information, and agricultural management practices. An individual who works with each farmer and each site forms the link between the biosolids generator, field operations, and the local communities.

Bio Gro Systems maintains and expands agricultural support for its land application programs by:

- Arranging field operational demonstrations,
- Sponsoring local programs (e.g., Ag Appreciation Days),
- Providing biosolids nutrient information to extension agents,
- Funding research and extension publications,
- Presenting information at sponsored breakfast and dinner meetings for the farming community, and
- Providing farmers with results of soil testing and field application reports.

Table 12-1. First year biosolids value.

Nutrient	Value per acre[†]
Nitrogen	$22.00
Phosphate ($P_2 0_5$)	10.35
Potash ($K_2 0$)	1.04
Sulfur	1.20
Magnesium	5.30
Manganese	0.18
Copper	0.15
Zinc	0.33
Lime	25.00
Spreading/tillage	25.00
First year savings	$90.55

[†]Based on a 1991 application of Blue Plains anaerobically digested biosolids for a 100-bu corn (*Zea mays* L.) crop.

Table 12-1 shows Bio Gro's method of providing farmers with actual cost savings (compared with current fertilizer prices) for realistic crop yield goals. By examining cost savings (rather than the total nutrient content) from biosolids, farmers can better estimate the value for their crop nutrient budgets.

COMMUNITY INVOLVEMENT

Working closely with farmers and their organizations, Bio Gro gains local support for bringing an urban material (biosolids) into surrounding rural communities. With the participation of farmers interested in receiving biosolids, our technical staff identifies and meets informally with key individuals in the local communities:

- County administrators and managers,
- County elected officials,
- State legislative representatives,
- Environmental organizations,
- Community associations, and
- Civic groups.

To increase public acceptance for land application, Bio Gro, with the support of the state regulatory agency and the biosolids generator, presents biosolids as a resource rather than a waste material. When local communities understand that biosolids are a beneficial end-use product resulting from the treatment of wastewater, they can perceive the link between wastewater treatment and biosolids utilization and the benefits derived from both.

COMMUNICATION CHANNELS

Public acceptance programs provide an on-going effort to maintain the flow of information and keep communication lines open among all interested parties. Concerns about health, odor, ground water contamination, declining property values, and other legitimate issues are anticipated and defused.

Bio Gro uses any or all of the following public information methods:

- Media contact: news releases and informational materials describing land application in general and project specifics;
- Written and audio visual materials: detailed land application manuals, specific project information, brochures, question and answer pamphlets, news reprints and slide presentations;
- Public meetings: public information, regulatory, and special interest groups;
- Tours: in the field and in concert with the treatment of wastewater by publicly owned treatment works (POTWs) to discharge clean effluent; and
- Regulatory liaison: to improve communication, Bio Gro provides regulators with technical and operational information beyond the minimum required for the permitting process. Regulators are also included in nonregulatory meetings, field days, tours and other community outreach programs.

For any biosolids management project, the effort to gain and maintain public acceptance is an integral part of the project throughout its lifetime. Different needs and concerns exist for different types of projects, but community acceptance can never be taken for granted. It is the strongest underpinning for any successful environmental project in our society today, and those who neglect outreach efforts risk failure, no matter how sound the project may be technically.

RESEARCH NEEDS

The type of information in Table 12-1, which forms the basis for agricultural value is based on organic N mineralization calculations developed in the 1970s. More up-to-date and more specific mineralization data is a significant need in order to maintain continued credibility for biosolids use as a nutrient source. Increasing emphasis on agricultural nutrient management will almost certainly affect land application in the near future. Realistic estimates of N availability under various soil and environmental conditions are essential for land application to remain a viable component of production agriculture.

Data on bioaccumulation of phosphorus (P) from biosolids and the pathways by which land applied biosolids impact sediment P are also a significant research need. Phosphorus forms from biosolids differ from fertilizer and animal waste sources, and the behavior of P-enriched sediment from these different sources have not been adequately researched.

Data on physical effects of land application operations (e.g., soil compaction and any concomitant problems associated with soil compaction) are also needed to better evaluate land application practices.

There is a critical need for data compilation and outreach for information on the issues relating to land application. Agricultural research in related areas should be evaluated with respect to its relevance to the issues surrounding land application. There is currently a general lack of awareness within the agricultural community of the research information and practices which form the basis of current biosolids programs. Efforts to gain new insights and data should be accompanied by a commitment to provide the results to the agricultural community and interested segments of the general public.

SECTION III

SPECIAL USES OF SEWAGE SLUDGE

13 Biosolids Utilization in Forest Lands

Charles L. Henry
Dale W. Cole

College of Forest Resources
University of Washington
Seattle, Washington

There are a number of reasons for considering forested sites as potential candidates for the use of biosolids:

1. Many forest soils have limited nutrients that can limit productivity. These nutrients are found in biosolids, especially nitrogen (N) and phosphorus (P).

2. Since relatively small amounts of food are gathered from a forested site in comparison to an agricultural site, many of the public health concerns should not be as critical as those associated with agricultural sites (Henry, 1989).

3. Forest soils usually have properties well suited to receive biosolids, including a great deal of organic carbon (C) that can store available N, a high infiltration rate that should minimize the potential for surface runoff, and a perennial root system that in some cases allow for year-round uptake of available nutrients.

4. The USA's extensive areas of forest land are potentially available for such a program.

POTENTIAL BENEFITS

Potential benefits from biosolids application to forested ecosystems can be classified in three categories: (i) soil improvement, (ii) increased timber production, and (iii) secondary benefits from understory vegetation response.

Sewage Sludge: Land Utilization and the Environment. SSSA Miscellaneous Publication.

Soil Improvement

The value of biosolids is its ability to amend the soil both by providing nutrients and by improving soil structural characteristics. Both short-term and long-term productivity can be improved by biosolids additions. There is an immediate supply of virtually every nutrient needed for plant growth in an available form, especially N. Additionally the fine particles and organics in biosolids can immediately and permanently enhance soil moisture and nutrient holding characteristics. In the long-term biosolids provide a continual slow release of nutrients as the organics decompose.

Growth Response

Growth response has been documented on a number of forest stands in Washington and in other states (Cole et al., 1986). Growth responses can be up to 100% for existing stands, and over 1000% for trees planted in soils amended with heavy applications of biosolids (Bledsoe, 1981; Henry & Cole, 1983; Henry, 1991). The magnitude of this response depends upon a number of site characteristics and stand ages (Henry et al., 1993).

Secondary Benefits

The third type of benefit from biosolids addition is more qualitative. Although immediately after application the site is greatly altered in appearance, within six months understory growth is often much more vigorous than before application. This is not only visually pleasing, but can be of commercial value to *brush pickers* harvesting ferns and other vegetation for floral arrangements. Increased understory is also typically higher in nutrients and can provide better habitat for wildlife. A number of wildlife studies have found increased populations of animals on sites receiving biosolids compared with nearby ones without.

DESIGN PHILOSOPHY

It is important to understand the characteristics of a forest application site in order to make a proper design. A site may require specific site studies to fine tune application rates based on nutrient transformations, uptake, and losses. Although biosolids have been shown to make trees grow faster, it also contains contaminants or levels of nutrients which, when not managed properly, could degrade the environment. Proper application of biosolids include consideration of the following management practices: (i) site evaluation criteria, (ii) nutrient loadings (primarily N), (iii) contaminant loadings, (iv) pathogen reduction, (v) slope restrictions, and (vi) buffer requirements.

Site Evaluation Criteria

Since the objectives of biosolids application to a forested site are to enhance tree growth, prevent environmental contamination and minimize operational costs, it is important that potential sites be evaluated with these objectives in mind. In particular, consideration must be given to physical factors of a site, such as: topography, soils and geology, vegetation (stand and understory characteristics), water resources, climate, and transportation and forest access.

Nutrient Management

Nitrogen has been traditionally considered the most important nutrient in determining biosolids application rates because it is needed as a soil supplement in much greater amounts than P or potassium (K). Typically the limiting constituent for land applications of biosolids is N, because when excess N is applied it often results in NO_3 leaching. A number of studies conducted in Washington confirmed this; heavy applications resulted in substantial increases of NO_3 in the ground water (Riekirk & Cole, 1976; Vogt et al., 1980).

Contaminant Loadings

Controlled loading that limits contaminants and proper management practices keep the risks associated with application of biosolids to forest lands exceptionally low. Contaminant loadings for forest applications are the same as those for agriculture (U.S. Environmental Protection Agency, 1993).

Pathogens

The U.S. Environmental Protection Agency (USEPA) has established two classifications for reducing the total number of pathogens in biosolids. Class A biosolids have no public access restrictions, whereas Class B biosolids require public access restriction for one year.

Slopes

Application of biosolids to excessive slopes will increase the risk of runoff from an application site. Typically, a slope limitation of 30% has been recommended.

Buffers

Buffers serve the following purposes: (i) to provide a factor of safety against oversprays or errors even when proper application and management techniques are used, and (ii) to absorb constituents and filter runoff from application areas. The condition of the ground surface is critical. In other words, bare soil will provide virtually no filtering, a grassed surface fair treatment, while a porous forest floor can provide excellent treatment and filtering. Buffer lengths are also dependent upon the type of waterway.

APPLICATION PROCEDURE

Three options have been developed for applications to forest lands: (i) biosolids applied to recent clearcuts, (ii) biosolids applied over the canopy of a young plantation, and (iii) biosolids applied under the canopy of an older stand (Henry & Cole, 1986).

Clearcuts

Clearcuts offer the easiest, most economical sites to apply biosolids. Since application will take place prior to tree planting, many agricultural biosolids application methods can be used. While clearcuts provide easier application, they also may require additional management practices to control grasses, rodents, and deer.

Young Plantations

Application of biosolids to existing stands are typically made by a tanker–sprayer system, which can apply 18% biosolids sprayed over-the-tree-canopy at 40 m into a plantation. This method requires application trails at 80 m intervals. A good criteria for the age–size is >5 yr or 1.2 to 1.5 m (4 to 5 ft) high. This minimizes maintenance suggested for clearcuts. Timing of applications is important with over–the–canopy applications. To aid in washing biosolids from the foliage and to keep biosolids off new foliage, spraying should take place during the rainy nongrowing season. Biosolids coating new foliage could retard the current years growth.

Older Stands

Applications to older stands have the advantage that biosolids can be applied year-round because spraying takes place under the foliage so that the foliage will not be affected. Application methods may be similar to those described for young plantations. Since stands are typically not in rows, however, some of the alternatives available for plantations may not work.

RESEARCH AND EDUCATIONAL NEEDS

Design Criteria for Application Rates, Buffers, and Slopes

The data base for appropriate design criteria needs to be broadened to include different soil conditions, tree species, type of biosolids, and climatic conditions. Many fine tuning efforts will be required before definitive application rate procedures are available.

Ecosystem Risk Assessment

Little research on the effects of biosolids on ecosystems has been conducted. This research needs to include effects on both animal and plant communities.

Economics

Costs of applications, as well as the benefits received through enhancement of forest productivity, must be determined.

Public and Regulatory Acceptance

Better communication is needed between researchers, municipalities and the general public.

REFERENCES

Bledsoe, C.S. 1981. Municipal sludge application to Pacific Northwest forest lands. Inst. Forest Resources Contrib. 41. College of Forest Resour., Univ. of Washington, Seattle.

Cole, D.W., C.L. Henry, and W. Nutter. 1986. The forest alternative for treatment and utilization of municipal and industrial wastewater and sludge. Univ. of Washington Press, Seattle.

Henry, C.L. 1989. Evaluation of comments on the proposed standards for management of sewage sludge: Non-agricultural land application. USEPA-NNEMS Publication. U.S. Environmental Protection Agency, Washington, DC.

Henry, C.L. 1991. Nitrogen dynamics of pulp and paper sludge to forest soils. Water Sci. Technol. 24:417–425.

Henry, C.L., and D.W. Cole. 1983. Use of dewatered sludge as an amendment for forest growth. Vol. IV. Inst. For. Resour. Univ. of Washington, Seattle.

Henry, C.L., and D.W. Cole. 1986. Pack Forest sludge demonstration program: History and current activities. p. 461–471. *In* D.W. Cole et al. (ed.) The forest alternative for treatment and utilization of municipal and industrial wastewater and sludge. Univ. of Washington Press, Seattle.

Henry, C.L., D.W. Cole, T.E. Hinckley, and R.B. Harrison. 1993. The use of municipal and pulp and paper sludges to increase production in forestry. J. Sustainable For. 1:41-55.

Riekirk, H., and D.W. Cole. 1976. Chemistry of soil and ground water solutions associated with sludge applications. p. 50-59. *In* R.L. Edmonds and D.W. Cole (ed.) Use of dewatered sludge as an amendment for forest growth. Vol. 1. Center for ecosystem studies. College of Forest Resour., Univ. of Washington, Seattle.

U.S. Environmental Protection Agency. 1993. Standards for the use or disposal of sewage sludge. (40 CFR Parts 257, 403, and 503). Final rate and phased-in submission of sewage sludge permit application (Revisions to 40 CFR Parts 122, 123, and 501). Federal Register 58(32): 9248–9415. U.S. Gov. Print. Office, Washington, DC.

Vogt, K.A., R.L. Edmonds, and D.J. Vogt. 1980. Regulation of nitrate levels in sludge, soil and ground water. p. 53-65. *In* R.L. Edmonds and D.W. Cole (ed.) Use of dewatered sludge as an amendment for forest growth. Vol. 3. Inst. Forest Resour. Univ. of Washington, Seattle.

14 Beneficial Effects Induced by Composted Biosolids in Horticultural Crops

Harry A. J. Hoitink

Department of Plant Pathology
Ohio Agricultural Research Center
The Ohio State University
Wooster, Ohio

Composts produced from biosolids generally induce beneficial effects after application to horticultural crops. They are used widely as peat substitutes in the production of nursery stock, during out planting in landscape settings and as top dressings on turf (Sanderson & Martin, 1974; Gouin & Walker, 1977; Logan et al., 1984; Gouin, 1985; Nelson & Craft, 1992). Significant quantities of essential nutrients are released by these composts (Chaney et al., 1980; Falahi-Ardakani et al., 1988). Crop response typically is positive (Logan et al., 1984; Falahi-Ardakani et al., 1988). In fact, it may be significantly above that which can be explained on the basis of rates of mineralization observed (Chen & Inbar, 1993; Inbar et al., 1993).

Suppression of diseases caused by plant pathogens and rhizosphere microorganisms deleterious to plant growth in part explains improved plant vigor (Hoitink et al., 1993). Other beneficial effects observed after compost applications include improved food quality (Vogtmann et al., 1993), increased numbers of flowerbuds on ornamentals (Logan et al., 1984) and improved survival of nursery stock after out planting. Physical properties of soils treated with composts typically also are improved. Infiltration and retention of water is enhanced and soil temperatures are decreased by compost applications (Dick & McCoy, 1993). A decrease in frost injury on mulched plants is reported frequently. This chapter emphasizes effects of composted biosolids on the severity of soilborne plant pathogens.

High temperatures (50 to 65°C) that occur during composting result in pathogen kill, as long as all parts of compost piles are exposed to adequate heat treatment. Mixing during the process ensures that this occurs. During the composting process, the organic fraction must be stabilized to a degree where fecal and plant pathogens can no longer utilize the compost as a food base after self-heating ceases. Some sludge composting systems cannot maintain temperatures predominantly below 65°C. These systems fail to stabilize

Sewage Sludge: Land Utilization and the Environment. SSSA Miscellaneous Publication.

composts because the microflora involved in composting is not adequately active at these high temperatures (Finstein & Hogan, 1993). The high available organic nutrient content (i.e., glucose), allows fecal and plant pathogens to recolonize such composts (Chen et al., 1988; Garcia et al., 1991; Bollen, 1993; Farrell, 1993). This causes problems during utilization. Allelopathy, which represents inhibition of one microorganism by another, is inhibited in nutrient-rich habitats such as poorly stabilized composts. For example, sclerotia of *Rhizoctonia solani* are not hyperparasitized by the biocontrol agent *Trichoderma* in fresh compost (Nelson et al., 1983). In mature compost, which is lower in cellulose content, this biocontrol agent hyperparasitizes *R. solani* and biocontrol prevails. The high glucose concentration in fresh compost is thought to repress chitinase activity required in biocontrol of this system (Chung et al., 1988). Thus, soluble organic C (glucose) in fresh organic matter, not only regulates the potential for regrowth by *Salmonella* (Burge et al., 1987), but also that for biological control of plant pathogens as well. This role of organic matter on gene expression in biocontrol agents needs to be explored further. Thus, composting systems must be able to provide optimum conditions for the process to avoid pathogen regrowth problems.

Biological control of plant pathogens induced by composts largely is due to the activity of biocontrol agents. This beneficial microflora, with the exception of heat resistant spores of *Bacillus* spp., is destroyed during the self heating process. This is less of a problem in composts produced in windrows on a forest floor than in enclosed (in-vessel) systems (Kuter et al., 1983). Thus, most biocontrol agents must recolonize composts after peak heating. This occurs naturally during curing of composts prepared from biosolids for biocontrol agents of some pathogens such as *Pythium* spp. and *Phytophthora* spp. (Kuter et al., 1988; Mandelbaum & Hadar, 1990; Hardy & Sivasithamparam, 1991). To avoid failures, specific biocontrol agent inoculants for pathogens such as *Rhizoctonia* have been developed that induce biological control predictably if this microflora is inoculated into compost after peak heating, but before substantial recolonization with mesophiles has occurred (Hoitink et al., 1993).

Carbon (C) availability, more specifically organic matter decomposition level, predicts the microbial species composition present in composts. In substrates high in stabilized biodegradable C such as composts and sphagnum peat with an H_2 (von Post scale) decomposition level, the microorganisms that predominate include fluorescent pseudonomads and other rhizobacteria that can function as biocontrol agents. These substrates are consistently suppressive to *Pythium* root rot (Boehm & Hoitink, 1992). In highly stabilized substrates, such as a sphagnum peat predictably conducive to *Pythium* root rot with an H_4 (von Post scale) decomposition level, the predominating microorganisms are *Arthrobacter, Micrococcus*, and other pleomorphic bacteria. This microflora cannot induce effective biological control (Boehm et al., 1993). These qualitative effects of organic matter decomposition level on microbial species composition and biocontrol need to be explored further.

The length of time that the disease suppression, induced by organic amendments, lasts depends on availability of C to the beneficial microflora after the amendment has been incorporated into soil. The rate of hydrolysis of fluorescein diacetate is an enzyme assay that predicts this effect in potting mixes and also in soil for some diseases (Boehm & Hoitink, 1992; Workneh, et al., 1993). Carbon availability in composts prepared from biosolids depends on the bulking agents used during the preparation of the compost. Although some quantitative information, utilizing cross polarization, magic angle spinning (CPMAS) ^{13}C-nuclear magnetic resonance (NMR) spectroscopy, has been developed on this lasting effect of composts, the length of time that the disease suppressive effects prevails remains largely undefined (Chen & Inbar, 1993; Hoitink et al., 1993).

Chemical properties of composts can affect diseases of crops in several ways. Composted biosolids typically cost less than peat, which results in application rates based on the peat substitute value rather than the fertility value of the compost. Thus, composts often are applied at loading rates exceeding those based on nutrient requirements for a given crop. It is not surprising that nurseries first using composted biosolids suffered increased losses from plant diseases that are aggravated by excessively high nitrogen (N) fertility regimes. Fireblight, *Phytophthora* stem dieback and some bacterial leaf spots are examples of diseases causing severe losses after excessive amounts of biosolids compost were applied (Hoitink & Fahy, 1986). Fusarium wilts respond to N in another way. They are enhanced by high NH_4 and low NO_3 nutrition regimes. Composted biosolids may enhance these diseases because the predominant form of N released is in ammonium form (Hoitink et al., 1993).

The salinity of composts is yet another chemical property that must be considered. High salinity enhances *Pythium* and *Phytophthora* root rots. The timing of compost application relative to planting must be such, therefore, that ample time is left for leaching before planting.

The foregoing reveals that opportunities for beneficial use of composted biosolids are significant. Great care must be taken that all aspects regulating fertility and crop response are accounted for in the overall management plan to avoid problems and reap all possible benefits. All too often, inadequate information is available to reach this goal.

RESEARCH NEEDS

Inadequate fundamental knowledge is available for predicting beneficial effects obtainable after compost applications. Interactions between soil organic matter decomposition level, populations and activity of biocontrol agents, soilborne plant pathogens and plant roots must be explored further. The composting process offers unique opportunities to study these phenomena, as well as other important biological soil processes. Nondestructive direct spectroscopic techniques such as CPMAS ^{13}C-NMR, Fourier transform infrared (FTIR), and diffuse reflectance infrared Fourier transform (DRIFT), must be applied to the analysis of soil organic matter in this application. This will improve the potential for predicting C availability and activity of pathogens and biocontrol agents, as well as nutrient release and other beneficial effects induced by composts prepared from biosolids.

REFERENCES

Boehm, M.J., and H.A.J. Hoitink. 1992. Sustenance of microbial activity in potting mixes and its impact on severity of *Pythium* root rot of poinsettia. Phytopathology 82:259–264.

Boehm, M.J., L.V. Madden, and H.A.J. Hoitink. 1993. Effect of organic matter decomposition level on bacterial species diversity and composition in relationship to *Pythium* damping-off severity. Applied Environ. Microbiol. 59:4171–4179.

Bollen, G.J. 1993. Factors involved in inactivation of plant pathogens during composting of crop resides. p. 301–318. *In* H.A.J. Hoitink and H.M. Keener (ed.) Science and engineering of composting: Design, environmental, microbiological and utilization aspects. Renaissance Publ., Worthington, OH.

Burge, W.E., P.D. Millner, N.K. Enkiri, and D. Hussong. 1987. Regrowth of salmonellae in composted sewage sludge. USEPA 600/2-86/106. (NTIS PB 87-129532/AS). Springfield, VA.

Chaney, R.L., J.B. Munns, and H.M. Cathey. 1980. Effectiveness of digested sewage sludge compost in supplying nutrients for soilless potting media. J. Am. Soc. Hortic. Sci. 105:485–492.

Chen, W., H.A.J. Hoitink, A.F. Schmitthenner, and O.H. Tuovinen. 1988. The role of microbial activity in suppression of damping-off caused by *Pythium ultimum*. Phytopathology 78:314–322.

Chen, Y., and Y. Inbar. 1993. Chemical and spectroscopical analyses of organic matter transformations during composting in relation to compost maturity. p. 551–600. *In* H.A.J. Hoitink and H.M. Keener (ed.) Science and engineering of composting: Design, environmental, microbiological and utilization aspects. Renaissance Publ., Worthington, OH.

Chung, Y.R., H.A.J. Hoitink, W.A. Dick, and L.J. Herr. 1988. Effects of organic matter decomposition level and cellulose amendment on the inoculum potential of *Rhizoctonia solani* in hardwood bark media. Phytopathology 78:836–840.

Dick, W.A., and E.L. McCoy. 1993. Enhancing soil fertility by addition of compost. p. 622–644. *In* H.A.J. Hoitink and H.M. Keener (ed.) Science and engineering of composting: Design, environmental, microbiological and utilization aspects. Renaissance Publ., Worthington, OH.

Falahi-Ardakani, A., J.C. Bouwkamp, F.R. Gouin, and R.L. Chaney. 1988. Growth response and mineral uptake of lettuce and tomato transplants grown in media amended with composted sewage sludge. J. Environ. Hortic. 6:130–132.

Farrell, J.B. 1993. Fecal pathogen control during composting. p. 282–300. *In* H.A.J. Hoitink and H.M. Keener (ed.) Science and engineering of composting: Design, environmental, microbiological and utilization aspects. Renaissance Publ., Worthington, OH.

Finstein, M.S., and J.A. Hogan. 1993. Integration of composting process microbiology, facility structure and decision-making. p. 1–23. *In* H.A.J. Hoitink and H.M. Keener (ed.) Science and engineering of composting:

Design, environmental, microbiological and utilization aspects. Renaissance Publ., Worthington, OH.

Garcia, C., T. Hernandez, and F. Costa. 1991. Changes in carbon fractions during composting and maturation of organic wastes. Environ. Manage. 15:433–439.

Gouin, F.R. 1985. Growth of hardy Chrysanthemums in containers of media amended with composted municipal sewage sludge. J. Environ. Hortic. 3:53–55.

Gouin, F.R., and J.M. Walker. 1977. Deciduous tree seedling response to nursery soil amended with composted sewage sludge. Hort. Science 12:45–47.

Hardy, G.E.St J., and K. Sivasithamparam. 1991. Sproangial responses do not reflect microbial suppression of *Phytophthora drechsleri* in composted eucalyptus bark mix. Soil Biol. Biochem. 23:757–765.

Hoitink, H.A.J., M.J. Boehm, and Y. Hadar. 1993. Mechanisms of suppression of soilborne plant pathogens in compost-amended substrates. p. 601–621. *In* H.A.J. Hoitink and H.M. Keener (ed.) Science and engineering of composting: Design, environmental, microbiological and utilization aspects. Renaissance Publ., Worthington, OH.

Inbar, Y., Y. Chen, and H.A.J. Hoitink. 1993. Properties for establishing standards for utilization of composts in container media. p. 668–694. *In* H.A.J. Hoitink and H.M. Keener (ed.) Science and engineering of composting: Design, environmental, microbiological and utilization aspects. Renaissance Publ., Worthington, OH.

Kuter, G.A., H.A.J. Hoitink, and W. Chen. 1988. Effects of municipal sludge compost curing time on suppression of Pythium and Rhizoctonia diseases of ornamental plants. Plant Dis. 72:751–756.

Kuter, G.A., E.B. Nelson, H.A.J. Hoitink, and L.V. Madden. 1983. Fungal populations in container media amended with composted hardwood bark suppressive and conducive to Rhizoctonia damping-off. Phytopathology 73:1450–1456.

Logan, T.J., W.R. Faber, and E.M. Smith. 1984. Use of composted sludge on different crops. Ohio Rep. 3:37–40.

Mandelbaum, R., and Y. Hadar. 1990. Effects of available carbon source on microbial activity and suppression of *Pythium aphanidermatum* and microbial activities in substrates containing composts. Biol. Wastes 26:261–274.

Nelson, E.B., and C.M. Craft. 1992. Suppression of dollar spot on creeping bentgrass and annual bluegrass turf with compost-amended topdressings. Plant Dis. 76:954–958.

Nelson, E.B., G.A. Kuter, and H.A.J. Hoitink. 1983. Effects of fungal antagonists and compost age on suppression of Rhizoctonia damping-off in container media amended with composted hardwood bark. Phytopathology 73:1457–1462.

Sanderson, K.C., and W.C. Martin, Jr. 1974. Performance of woody ornamentals in municipal compost media under nine fertilizer regimes. Hort. Science 9:242–243.

Vogtmann, H., K. Matthies, B. Kehres, and A. Meier-Ploeger. 1993. Enhanced food quality induced by compost applications. p. 645–667. *In* H.A.J. Hoitink and H.M. Keener (ed.) Science and engineering of composting: Design, environmental, microbiological and utilization aspects. Renaissance Publ., Worthington, OH.

Workneh, F., A.C.H. Van Bruggen, L.E. Drinkwater, and C. Shennan. 1993. Variables associated with corky root and Phytophthora root rot of tomatoes in organic and conventional farms. Phytopathology 83:581–589.

15 Use of Sewage Sludge on Park and Recreational Lands

Antonio J. Palazzo
Iskandar K. Iskandar

U.S. Army Cold Regions Research and Engineering Laboratory
Hanover, New Hampshire

One of the primary factors in selecting potential sites for land application of sewage sludge is subsequent land use. Park and recreational areas, comprising many acres of land in the USA have a wide array of uses. The type of use dictates the vegetation management they receive, ranging from intensive to intermediate for golf courses and recreational lands to very low for park areas. The intensive management conducted on areas such as golf courses, where the predominant vegetation is turfgrasses, usually includes mowing every other day and fertilizing and irrigating periodically. These areas require fertile soils and daily inspection by the golf course superintendent to obtain the type of cover required. Areas with intermediate management include ballfields and parks that have turf or woody vegetation. Management at these sites is usually less intensive, primarily due to the lack of funding from municipalities. Public lands with very low management usually include highway medians and wildlife areas. In these areas, management techniques such as fertilization and mowing are very limited. Regardless of their use, these areas are managed much differently than private agricultural lands where sludge is usually applied.

The geographical and political characteristics of public use areas make them candidates for land application of sludge. They are large, remote from residential areas, and the vegetation growing on them is not for human consumption. Sludge use would also be beneficial because of the agronomic requirements for the growth of vegetation. With the exception of golf courses, public lands usually have poor vegetation growth due to deficiencies in plant nutrients, low soil pH in humid areas, and poor soil structure. Sewage sludge can help satisfy these nutrient requirements and compensate for lack of funding to purchase soil amendments.

Sewage sludge has been used on golf courses for many years. Spindler (1986) reported that Milorganite®, a dried sewage sludge product from the city of Milwaukee, WI, has been in use for nearly 60 yr. As of 1986, 3 295 430 Mg

Sewage Sludge: Land Utilization and the Environment. SSSA Miscellaneous Publication.

of Milorganite® sludge had been sold. In fact, Milorganite® was in use prior to the more complete fertilizers, such as 10-10-10 grade fertilizer, that are commonly used today. The benefits of using sludge as fertilizer and soil amendment on turf has been reported by Murray et al. (1979) and others. On golf courses these benefits include increased soil fertility, pH, soil water retention, microbial activity, decreased thatch buildup, improvements in soil structure, lower nematode populations, and darker turf color. Similar benefits are also noted for sports turf, with the added advantage that the resulting well-maintained, sludge-fertilizer turf reduces injuries to athletes (Alexander, 1991). The extra organic matter supplied through the sludge also helps to alleviate soil compaction problems when growing turf in more intensely trafficked areas. From reviewing the literature, it appears that in the early 1980s there was some concern that Milorganite® might be the cause of Lou Gehrig's disease, but this was not found to be true (Anonymous, 1987).

Parklands receive less intensive use and management than golf courses and ballfields. One application of sewage sludge when the herbaceous vegetation is sown is usually all that is required to obtain a uniform, dense turf cover at low cost. One application of sewage sludge at the rate of 100 Mg ha^{-1} was sufficient to revegetate highly acidic soils (pH 2.5) on a federally owned waterway in Delaware for >15 yr (Palazzo et al., 1994). The vegetation response to the sludge helped to reduce soil erosion and soil acidity, promote wildlife, and reduce acidic runoff into an adjacent waterway (U.S. Army Corps of Engineers). The use of composted sludge to improve soils for plant growth has been reported for the communities of Fairfield, CT; St. Paul, MN; Seattle, WA; and Ontario, Canada (U.S. Environmental Protection Agency, 1973; Faust & Romano, 1978). In Fairfield and Ontario the sludge was used to help establish vegetation on closed landfills that will be converted to parklands.

Wakefield and Sawyer (1986) reported that sewage sludge was beneficial for the establishment and maintenance of trees and turf along roadsides. The benefits described are similar to those stated above and include increased soil moisture holding capacity, soil nutrients, and soil organic matter.

Public lands may also be used by communities as test or demonstration sites to evaluate the benefits of sludge prior to its application on local private property. In Monroe County, New York, and Florence, AL, publicly owned land was used as a test site prior to applying municipal sludge to farmland (National Association of Conservation Districts, 1982).

In most cases where sludge is used, both the supplier and user of sludge receives benefits. The supplier or municipality will benefit by finding an outlet for this material, and the user will receive a cost-effective nutrient source or soil conditioner. Questions do remain, however, about its application to land, mostly centering on possible health and environmental effects due to pathogens, odors, and movement of pollutants. Most of these concerns can be addressed by involving all parties during the planning and permitting process and through the proper treatment of the sludge (National Association of Conservation Districts, 1982). Good communication among all parties involved is mandatory to avoid administrative problems during the approval process.

Research is needed to develop more fundamental information on the use of sewage sludge on park and recreational lands. The current literature is mostly limited to case studies on the use of sludge by a municipality. This information is useful, but well designed studies are required to generate more information to design future applications. To satisfy these needs, research studies should evaluate both the establishment and long-term agronomic and cost benefits of using sludge to maintain vegetation and promote soil fertility. The results of these studies may then be used to develop guidelines for establishing and managing sites which have received sludge so that the soil amendment benefits may be clearly evaluated and compared to more conventional management techniques.

REFERENCES

Alexander, R. 1991. Sludge compost: Can it make athletic fields more playable? Lawn Landscape Maint. 12:46–49.

Anonymous. 1987. EPA, scientists call Milorganite safe to use. Landscape Manage. 26:12.

Faust, J., and L.S. Romano. 1978. Composting sewage sludge by means of forced aeration at Windsor, Ontario. Ontario Ministry of Environment Sludge Utilization and Disposal Conference, Toronto, Canada. 20–21 Feb. 1978. Ontario Ministry of Environment, Ottawa, Canada.

Murray, J.J., S.B. Hornick, and J.C. Patterson. 1979. Use of sewage sludge in the vegetation of urban lands. Proc. of the Second Conference on scientific research in the National Parks. San Francisco, CA. 26-30 Nov. 1979. National Park Service, San Francisco, CA.

National Association of Conservation Districts. 1982. Sludge and the land: The role of soil and water conservation districts in land application of sewage sludge. USEPA Rep. 430/9-82-007. U.S. Gov. Print Office, Washington, DC.

Palazzo, A.J., C.R. Lee, and R. Price. 1994. Long-term plant persistence on highly acidic soils amended with organic materials in two climatic zones. Proc. of the Third Int. Symp. on Plant-Soil Interactions. Brisbane, Australia, 12-16 Sept. 1993. (In press).

Spindler, J. 1986. The Milorganite story. The grass roots. p. 1-5. Wisconsin Golf Course Superintendents Assoc. May/June Issue. Madison, WI.

U.S. Army Corps of Engineers. Transformation of lands at the Chesapeake and Delaware Canal. Philadelphia District, Philadelphia, PA.

U.S. Environmental Protection Agency. 1973. Park development with wet digested sludge. USEPA Rep. EPA-R2-73-143. U.S. Gov. Print. Office, Washington, DC.

Wakefield, R.C., and C.D. Sawyer. 1986. Use of composted sewage sludge on roadside vegetation. Univ. of Rhode Island Agric. Exp. Stn. Contrib. 2353. Kingston.

16 Use of Sewage Sludge for Land Reclamation in the Central Appalachians

W. Lee Daniels
Kathryn C. Haering

Department of Crop and Soil Environmental Sciences
Virginia Polytechnic Institute and State University
Blacksburg, Virginia

Sewage sludge has been used by various researchers as a soil amendment in surface mine reclamation with excellent results (Sopper, 1992). Application and monitoring protocols for the use of sewage biosolids on mined lands have been developed and tested in western Pennsylvania (Sopper & Seaker, 1983), Illinois (Peterson et al., 1982), and at several other areas across the USA.

Sewage sludge, however, has generally not been utilized for reclamation in the central Appalachian coal mining region because environmentally sound guidelines for its use in the region have been lacking, and a large-scale demonstration of its effectiveness had not been implemented before this program.

Over the past 12 yr we have been involved in a progression of field-oriented biosolids research and demonstration programs at the Powell River Project Research Area in Wise County, Virginia. The Powell River Project site was chosen for a number of reasons. First, there have been numerous detailed studies on the mine soils, water quality, and land resources within this area over a long period of time (Daniels & Amos, 198la, 1981b, 1982; Roberts et al., 1988a, 1988b, 1988c; Moss et al., 1989). The overburden and mine soils in this area are very productive compared with the rest of the Appalachian region, and are generally nonacidic. Second, due to the active mining operation in the area, we had good surface runoff control and historical baseline water quality data from sediment ponds, wells, and streams surrounding the site. Third, we had been involved in a cooperative monitoring program at the active mining site, and knew the exact composition and characteristics of the spoils across the site.

In 1982, we installed a small plot experiment to evaluate the potential of municipal sewage sludge as a mine spoil amendment to reconstruct topsoil substitutes from hard rock overburden. Tall fescue (*Festuca arundinacea* L.) standing biomass was increased dramatically by application of biosolids up to loading rates of 112 Mg ha^{-1}, but increasing the loading rate to 224 Mg ha^{-1} had no further yield effect (Roberts et al., 1988a, 1988c). Heavy metal concentrations in fescue tissue increased along with the loading rate in this

Sewage Sludge: Land Utilization and the Environment. SSSA Miscellaneous Publication.

experiment, but did not approach established forage toxicity standards (Roberts et al., 1988c). Pine (*Pinus* sp.) seedlings planted in split plots in the same experiment exhibited a strong negative response to increased biosolids loading rate, presumably due to the high pH and salt loadings associated with the biosolids application (Moss et al., 1989). We continue to monitor this experiment today, and the relative effects of the biosolids treatment on forage production are still striking and consistent after 12 yr.

Once we had completed the intensive 5-yr plot study discussed above, we decided that it was appropriate to move to a full-scale trial of biosolids in southwest Virginia. In association with Enviro-Gro Technologies, we developed a cooperative research and demonstration project using a mixture of sludge cake and composted wood chips (*mine mix*) to reclaim a recently regraded surface mine. The overall objective of this study was to evaluate the logistical, economic, and environmental feasibility of biosolids utilization on southwest Virginia mined lands through the development and monitoring of an operational-scale pilot program in the Powell River Project research watershed.

Specifically, we planned to carefully monitor the effect of biosolids application on forage quality, soil properties, and long-term surface and ground water quality around the utilization area. Loading rate plot studies were nested within the treatment area to enable us to determine optimal application rates to insure reclamation success. This project was cooperative among Virginia Tech, Enviro-Gro Technologies, the Virginia Center for Innovative Technology, and the Powell River Project.

SITE SELECTION AND COOPERATORS

An active surface mining operation (Red River Coal Co.) within the Powell River Project Research Area agreed to provide ≈70 ha of newly graded mine soils for the project (Fig. 16-1). The site was originally mined in the mid-1970s and was dominated by narrow benches, highwalls, and steep outslopes. Numerous areas of acidic spoil were exposed at the surface and the overall topography was quite rough. The site was remined between 1987 and 1989 and is now relatively flat and supports a highly productive hayland–pasture land use. The full approval of the land owner (Penn Virginia Resources Corp.), the mining company (Red River Coal Co.), and the various regulatory agencies involved was required for such a venture to be successful. Penn Virginia Corp. agreed in principle to the project, Norfolk-Southern Corp. agreed to haul the mine mix for a reasonable rate, and Enviro-Gro Technologies agreed to supply and handle the sludge materials. In April 1989, the Virginia Division of Mined Land Reclamation approved a permit revision by Red River Coal Co. to substitute a compost–sewage sludge mixture for normal fertilization. This permit revision was also reviewed by the Virginia Water Control Board and the Virginia Department of Health.

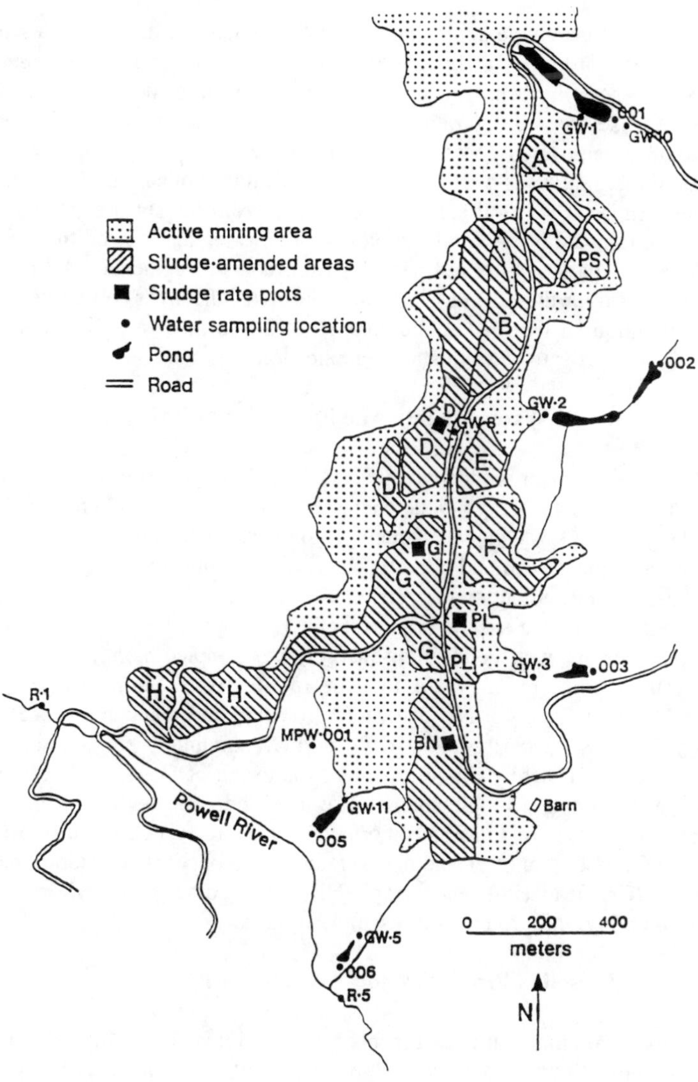

Fig. 16-1. Map of biosolids utilization area at the Powell River Project field site in Wise County, Virginia.

MATERIALS AND METHODS

Mine Soil Mapping, Sampling, and Site Preparation

Before sewage sludge application in 1989, we mapped the area at a scale of 1:3000 on standard black and white aerial photos. Contrasting mine soils were delineated based on rock type composition, slope class, depth to hard rock, wetness, and surface configuration. The maps were then used to calculate the amount of suitable area across the site for biosolids application. Large bulk samples of the mine soils were taken before biosolids application on a 50 m grid across the entire area. Soil samples were taken from the surface (0 to 15 cm) layer of the mine soils. All soil samples were air dried and passed through a 2-mm sieve to separate soil-sized particles from coarse fragments. All analyses were performed on the <2-mm fraction. Soil pH of the areas to be treated was measured and areas that had acidic material at the surface were limed at a rate of 22 Mg ha^{-1} prior to sewage sludge application.

Water Quality Monitoring Program

One of the major concerns associated with large-scale sewage sludge utilization programs is the potential contamination of surface and ground waters, particularly by NO_3. Our water quality monitoring program was cooperative with, and in some cases in addition to, the regular monitoring program that Red River Coal Co. carried out on the permit. Sediment ponds were sampled weekly to bi-weekly during the first year and monthly during the second year. Ground water (piezometers, wells, and fill discharges) was sampled monthly and surface water instream sites monthly. A total of 15 sites were sampled (Fig. 16-1). Samples were analyzed for NO_3, pH, total iron (Fe) and manganese (Mn), total suspended solids (TSS), conductivity, acidity, alkalinity, sulfate, and total dissolved solids (TDS) by Environmental Monitoring of Coeburn, VA. Before analysis, each sample was split and half was acidified. The acidified sample was analyzed in our lab for total phosphorus (P), calcium (Ca), magnesium (Mg), aluminum (Al), sodium (Na), and metals [zinc (Zn), copper (Cu), cadmium (Cd), chromium (Cr), lead (Pb), and nickel (Ni)]. We composited sediment pond samples monthly and ground water samples quarterly.

Biosolids Properties and Application Procedures

The biosolids material used for this project originated in Philadelphia, and was a mixture of composted wood chips and stabilized anaerobic sludge cake. The composted wood chips and the sludge cake are mixed 50:50 by volume, but the actual spread material is approximately two-thirds composted wood chips and one-third sludge cake on a dry weight basis. All heavy metal loading rates for the material were less than the Virginia lifetime loading limits (at that time) for soils with a cation-exchange capacity of less than 5 $cmol_c$ kg^{-1} (Table 16-1). The amount of total N added with the biosolids was quite high (≈4400 kg ha^{-1}), but the C/N ratio of the material was above 25:1, so major NO_3 leaching was not

Table 16-1. Approximate nutrient and metal loading rates at *mine mix* application rate of 368 Mg ha^{-1}.

Element	Added†	Element	Added†
	kg ha^{-1}		kg ha^{-1}
C	121 200	Mn	167
N	4400	Zn	207
P	2550	Cu	108
K	1100	Pb	58
Ca	6050	Ni	27
Mg	2020	Cd	1

†kg ha^{-1} = lb $acre^{-1}$ x 1.12.

anticipated.

The sludge–wood chip mixture was hauled to the site via rail and covered truck, and was applied to the 70-ha site during the summers of 1989 and 1990. The *mine mix* was applied by either spreader truck or with a spreader unit pulled behind a farm tractor. The wet loading rate of the mix was ≈780 Mg ha^{-1} (350 ton $acre^{-1}$). The depth of the material was ≈12 cm before deep incorporation. The *mine mix* materials were incorporated by chisel plowing (20 cm) and heavy disking immediately after spreading, and were seeded immediately to a mixed grass–legume pasture stand. Biosolids were not spread within 5 m of outslopes, drainage ditches, and sediment ponds, or in designated wetland areas. We subsampled the materials just before application for analysis and verification of total loading rates. Our actual dry loading rate was ≈368 Mg ha^{-1} *mine mix*, or 100 Mg ha^{-1} (45 ton $acre^{-1}$) as dry sewage sludge cake. Three sets of sewage sludge rate plots were laid out within the area to be treated. Each set consisted of six 5 by 30 m plots separated by 3 m alleys. In these plots, *mine mix* was applied at rates of 92 Mg ha^{-1} (1/4 basic rate), 184 Mg ha^{-1} (1/2 basic rate), 368 Mg ha^{-1} (basic rate), and 552 Mg ha^{-1} (1-1/2 basic rate). At these rates, we were adding 25, 50, 100, and 150 Mg ha^{-1} sewage sludge cake, respectively. We also included both a fertilized and non-fertilized control. These plots were spread during the summer of 1989 and seeded with the same forage mixture used for the larger treatment area.

Soil Chemical Analyses

A total of 84 soil samples were taken from the 40-ha area treated in 1989 before biosolids application. Three soil samples were also taken from each of the loading rate plots. In 1990, 37 samples were taken from the 18-ha area before biosolids were applied. These samples were analyzed for soil pH and double-acid extractable P, K, Ca, Mg, Zn, and Mn. A randomly determined subset of 32 samples from the areas treated in 1989, along with all 37 samples taken from the area treated in 1990, and all of the loading rate plot samples were analyzed for total C, total N, bicarbonate-extractable P, exchangeable bases, exchangeable Al and acidity, and effective cation-exchange capacity (ECEC). During 1990, 32 samples were taken from the areas where biosolids had been applied in 1989, and analyzed for the same parameters as the preceding year's samples. Standard methods of analysis for the soils were used as specified by Haering and Daniels (1991).

Plant Tissue Sampling and Chemical Analyses

Samples of plant tissue were taken by hand clipping to ground level. Forage yields in the large sludge-amended areas were determined by clipping 8 to 10 1 m^2 quadrates in randomly selected areas. In the loading rate plots, two 0.5 m^2 quadrates in each plot were randomly selected for fescue tissue sampling. The tissue was oven dried at 65°C for 48 hours and weighed to determine dry weight yield. Plant tissue samples taken from the rate plots were ground for use in determining nutrient and metal levels. Dried *mine mix* and dried plant tissue were analyzed by standard methods as discussed by Haering and Daniels (1991).

RESULTS AND DISCUSSION

Mine Soil Mapping and Control of Application Areas

Accurate and detailed mine soil mapping is critical to the safe and successful use of sewage sludge materials on mined lands. We found that the mapping criteria and scale used were quite satisfactory for delineating contrasting spoil types, hot spots, and areas that should not be treated such as wetlands, drains, and zones of instability. The compiled maps were analyzed with a digital planimeter to estimate the total acreage to be treated, and the tonnage hauled to the site was then calculated from those estimates. The maps were also used during spreading operations to insure that only appropriate areas of the mine were treated, and to record treatment progress. Monitoring points, stockpiles, and other important management features can also be added to the mapping base once it is developed. We do not recommend the treatment of any mined area without detailed site-specific mapping.

Mine Soil Properties Before and After Biosolids Application

The application of the sludge–wood chip mixture raised the average pH of the mined area treated during 1989 from 5.6 to 6.5 and increased soil nutrient levels, C content, and exchangeable bases (Table 16-2). Soil Zn content increased greatly but remained well below levels that would be toxic to metal-sensitive vegetation. In the biosolids loading rate plots, pH, soil nutrient levels, C content, and metal (except Mn) levels increased with increasing biosolids rate (Table 16-3). In plots where the *mine mix* was applied at 184 Mg ha^{-1} or more, all nutrient levels were higher than those in the fertilized control.

Table 16-2. Soil pH, C, N, bicarbonate-extractable P, double-acid extractable nutrients, and exchangeable cation levels in surface layer of large sludge-amended areas before (1989) and after (1990) sludge application (n = 32).

Time of sampling	pH	C	N	P[†]	Double-acid extractable P	K	Ca	Mg	Zn	Mn
	mg kg^{-1}				mg kg^{-1}					
Before application	5.6b[‡]	20.9b	2.1b	6b	33b	62b	615b	209b	3b	50a
After application	6.5a	80.1a	5.2a	150a	310a	14la	2207a	380a	75a	59a

Time of sampling	Exchangeable Ca	Mg	K	Al	Acidity	ECEC[§]
	cmol$_c$ kg^{-1}					
Before application	3.02b[‡]	1.78b	O.17b	0.50a	4.79b	5.47b
After application	10.27a	3.82a	0.65a	O.16a	11.38a	14.91a

[†]Bicarbonate-extractable.
[‡]Means followed by different letters by column are significantly different, P = O.05 (Two sample t-test).
[§]ECEC, effective cation-exchange capacity.

Table 16-3. Soil pH, C, N, bicarbonate-extractable P, double-acid extractable nutrients, and exchangeable cation levels in surface layer of sludge rate plots before (1989) and after (1990) treatment application.

Treatment	Rate	pH	C	N	P†	Double-acid extractable					
						P	K	Ca	Mg	Zn	Mn
	Mg ha^{-1}	mg kg^{-1}				mg kg^{-1}					
		Before application									
Control	--	5.5a‡	20.7a	1.0a	3a	29ab	60a	598a	202a	3a	56a
Fertilized	--	5.2a	23.8a	1.3a	3a	30ab	58a	547a	210a	2a	48a
Sludge	92	5.6a	24.4a	1.0a	5a	37ab	62a	610a	203a	3a	52a
Sludge	184	5.3a	19.5a	1.4a	3a	26b	58a	567a	214a	2a	44a
Sludge	368	5.6a	20.1a	1.4a	4a	31ab	63a	610a	202a	3a	60a
Sludge	552	5.7a	20.8a	1.4a	5a	43a	61a	642a	231a	3a	51a
		After application									
Control	--	5.3c‡	22.0c	2.5c	10e	29e	82c	512d	193d	3e	38b
Fertilized	--	5.4c	24.3c	2.6c	32de	75d	89c	662d	293c	6e	49ab
Sludge	92	6.1b	36.9c	2.9c	43d	150c	92c	1074c	273c	25d	50ab
Sludge	184	6.5ab	60.5b	4.7b	100c	274b	128b	1812b	343b	56c	59a
Sludge	368	6.6ab	128.3a	7.9a	158b	352a	140b	2579a	402ab	88b	40b
Sludge	552	6.7a	142.5a	9.2a	215a	370a	204a	2906a	457a	108a	43b

continued

Table 16-3. Continued.

Treatment	Rate	Exchangeable Ca	Mg	K	Al	Acidity	ECEC[§]
	$Mg\ ha^{-1}$	.... $cmol_c\ kg^{-1}$					
		Before application					
Control	--	3.15a[‡]	1.90ab	0.17a	0.63a	4.73ab	5.85a
Fertilized	--	2.78a	1.97ab	0.17a	0.45ab	5.13ab	5.37a
Sludge	92	2.86a	1.78b	0.17a	0.42ab	4.07ab	5.23a
Sludge	184	2.93a	2.00ab	0.17a	0.34ab	5.82a	5.44a
Sludge	368	3.07a	1.88ab	0.18a	0.46ab	3.38b	5.59a
Sludge	552	3.15a	2.08a	0.17a	0.11b	4.80ab	5.51a
		After application					
Control	--	2.08c[‡]	1.52e	0.24d	1.01a	5.90d	4.84c
Fertilized	--	5.37bc	1.83de	0.29d	0.61ab	6.11d	8.09bc
Sludge	92	4.58bc	2.23d	0.29d	0.12b	5.20d	7.22c
Sludge	184	7.54b	3.03c	0.50c	0.13b	8.87c	11.21b
Sludge	368	13.02a	4.39b	0.79b	0.12b	13.47b	18.32a
Sludge	552	14.52a	5.16a	1.23a	0.12b	16.15a	21.02a

[†]Bicarbonate-extractable.

[‡]Means followed by different letters by column and time of sampling are significantly different, P = O.05 (Fisher's LSD).

[§]ECEC, effective cation-exchange capacity.

Carbon (C), N, bicarbonate-extractable P, and extractable K levels in the 92 Mg ha^{-1} plots, however, were not significantly different from the levels in the fertilized plots. Exchangeable cations (except Al) and ECEC also increased with the loading rate (Table 16-3). Exchangeable Al appeared to decrease as biosolids rate increased; this may be a result of the complexation of Al with organic matter.

Forage Yield and Quality

Average first-year forage yields across the treated mined area were 9.5 Mg ha^{-1}, while second-year yields averaged 7.3 Mg ha^{-1}. Forage yields in the rate plots were somewhat lower than those in the larger mined areas, perhaps because the plots were seeded later in the year and because some of the mined area already had existing vegetation at the time of spreading. In the biosolids loading rate plots, forage yields generally increased with the loading rate (Fig. 16-2). First-year yields on sludge-amended plots were significantly higher than the unfertilized control, but statistically equal to the control plots which received conventional fertilization. After two growing seasons, however, forage yields on the three highest application rates were significantly higher than both the fertilized and unfertilized controls. These trends have become more apparent since 1990, and by the fall of 1993 the biosolids-treated plots supported a much more vigorous and diverse stand of vegetation than the fertilized control plots.

Plant tissue analysis over three years indicated that tissue Zn and Cu content tended to increase with loading rate, although Zn and Cu levels were well below potentially toxic concentrations (Table 16-4). Tissue Mn and Fe levels were not significantly affected by sludge rate. Fescue tissue Pb, Cr, Ni, and Cd levels also showed no consistent biosolids treatment effect when compared with untreated controls. This may be due to yield dilution effects in the higher producing biosolids treatments, but could also reflect background metal uptake from the fresh, unweathered spoil materials.

Water Quality

Results of the intensive water quality monitoring program associated with this project are detailed by Haering and Daniels (1991, 1992), and generally show negligible effects of the land application practice. Water quality was monitored at 15 locations across the site (sediment ponds, wells, piezometers, fill discharges, and two surface instream sampling locations in the Powell River). Nitrate and heavy metals were the major parameters of concern in the water quality monitoring program.

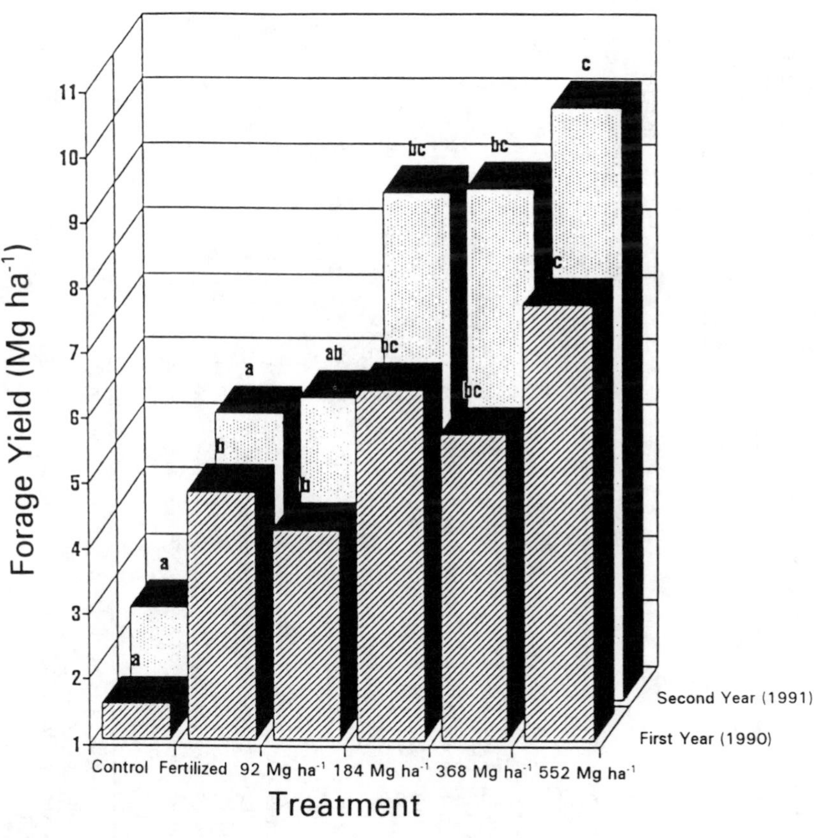

Fig. 16-2. Standing biomass from biosolids loading rate experiment measured in the fall of 1990 and 1991.

Table 16-4. Metal levels in forage harvested from sludge rate plots in 1990, 1991, and 1992 (1, 2, and 3 yr after sludge application).

Treatment		Rate	Metal †							
			Zn	Mn	Cu	Fe	Pb	Cr	Ni	Cd
		Mg ha^{-1}	mg kg^{-1}							
1990	Control	--	38.5c‡	489a	12.1c	338b	1.0b	0.9a	5.7a	na§
	Fertilized	--	38.9c	426a	14.8c	534a	1.4b	0.5a	1.6b	na
	Sludge	92	48.9bc	283abc	13.8c	396ab	1.lb	0.4a	1.3b	na
	Sludge	184	63.3ab	217bc	18.0bc	468ab	1.4b	0.6a	0.7b	na
	Sludge	368	65.6a	178c	24.8a	503ab	2.1a	0.8a	0.6b	na
	Sludge	552	70.5a	271abc	21.3ab	417ab	1.2b	0.6a	0.6b	na
1991	Control	--	36.6a	161c	15.6abc	412a	1.6a	0.4ab	1.4b	0.28b
	Fertilized	--	41.5a	190c	19.6a	328a	1.5a	0.5a	0.8b	0.31ab
	Sludge	92	40.8a	354bc	17.6ab	414a	1.6a	0.5a	2.1b	0.28b
	Sludge	184	46.8a	276bc	16.0abc	357a	1.2a	0.3b	1.8b	0.30b
	Sludge	368	56.1a	655a	14.1bc	355a	1.5a	0.2b	6.1a	0.38a
	Sludge	552	50.9a	425b	12.9c	349a	1.5a	0.4ab	3.4b	0.29b
1992	Control	--	33.0ab	630a	12.9a	443a	na§	0.7a	6.8a	0.46ab
	Fertilized	--	29.0b	274b	12.5a	447a	na	0.6a	2.7b	0.36b
	Sludge	92	39.2ab	313b	11.9ab	269b	na	0.6a	1.6b	0.44ab
	Sludge	184	40.1a	99b	10.1b	285b	na	0.7a	0.9b	0.67ab
	Sludge	368	36.6ab	139b	11.3ab	262b	na	0.6a	0.8b	0.42ab
	Sludge	552	41.2a	92b	12.6a	244b	na	0.6a	0.7b	0.72a

†Values were determined by inductively coupled plasma spectrometry (Zn, Mn, Cu, and Fe) and flameless atomic absorption spectrometry (Pb, Cr, Ni, and Cd).

‡Means followed by different letters by column within years are significantly different, P = O.05 (Fisher's LSD).

§na = not analyzed.

Nitrate

Since the spoil materials used to construct the mine soils contain trace amounts of NO_3, relatively high levels of NO_3 (up to 5 mg L^{-1}) were measured before application. The NO_3 levels in sediment ponds and valley fill discharge points showed no evidence of being influenced by biosolids application (Table 16-5). Nitrate levels in two piezometers, however, may have increased as a result of biosolids application. In one piezometer, NO_3 concentration increased to near 6 mg L^{-1} during the winter of 1990–1991, but dropped within 1 yr to below pre-application levels. Nitrate levels in the other piezometer have been consistently high (often approaching 10 mg L^{-1}) throughout the study. This piezometer dries up periodically, so no pre-application baseline data was collected. Nitrate levels have not increased consistently in any of the other ground water sampling locations, so the elevated NO_3 levels in this particular piezometer may be a result of surface water charging.

Heavy Metals

The risk of significant heavy metal contamination of ground water as a result of sewage sludge land application is generally thought to be minimal. We measured the heavy metal content of both sediment ponds and ground water, and found that Cd and Pb were nondetectable. Total Zn and Cu levels in both ponds and ground water were very low, and did not appear to increase as a result of biosolids application, except in one piezometer. The Zn concentration in this piezometer during the first quarter of 1990 was relatively high (1.91 mg L^{-1}), although still well below the drinking water standard of 5 mg L^{-1}. Since the method we used for metal determination also measures metals adsorbed on sediment and the water collected from this well is often turbid, we feel that this is a reflection of Zn bound on the sediment rather than in the ground water itself. This further reinforces our conclusion regarding surface water recharging at this particular location that also exhibited the high NO_3 values discussed earlier. Trace levels of both Cr and Ni were detected in several ponds and ground water sampling points, but similar concentrations of these elements were measured both before and after biosolids application indicating that they are probably present in the native spoil. Both Cr and Ni were undetectable in samples taken from most of the water sampling points.

Table 16-5. Average NO_3 levels in sediment pond discharge from the Powell River sludge application site.

Sampling date[†]		Sediment pond				
		001	002	003	005	006
		mg L^{-1}				
		Before application				
April	1989	1.11	2.04	1.73	3.96	1.48
May	1989	1.48	1.75	1.27	3.18	1.85
		After application				
June	1989	1.96	2.19	0.94	4.24	0.71
July	1989	1.69	1.70	1.06	--[‡]	--
August	1989	0.81	1.73	1.71	--	--
September	1989	0.75	1.24	1.18	--	0.71
October	1989	0.93	4.34	1.26	--	0.80
November	1989	0.65	4.08	0.91	1.84	0.64
December	1989	2.67	4.40	1.33	3.18	0.49
January	1990	1.72	3.77	1.74	3.28	0.73
February	1990	1.49	2.37	1.24	1.74	0.72
March	1990	0.91	2.14	0.64	2.50	0.57
April	1990	0.83	3.06	1.08	2.54	0.63
May	1990	0.51	1.91	0.98	2.04	0.35
June	1990	0.43	1.94	1.34	1.86	1.38
July	1990	0.39	1.21	1.27	1.58	0.12
August	1990	0.39	1.14	1.61	1.24	--
September	1990	0.71	0.82	2.77	2.27	--
October	1990	1.14	0.91	0.88	2.53	--
November	1990	0.77	1.04	0.81	2.47	--
December	1990	1.07	1.36	1.10	2.86	1.02
January	1991	1.77	--	1.93	4.21	1.68
February	1991	1.00	--	1.37	3.22	1.90
March	1991	na[§]	--	1.54	na	2.93
April	1991	1.46	--	1.27	2.58	2.55
May	1991	1.55	--	1.92	2.32	--
June	1991	0.57	--	0.64	0.73	--
3rd quarter/1991		1.03	--	1.36	3.41	--
4th quarter/1991		1.88	--	0.54	1.19	0.41
1st quarter/1992		2.44	--	0.82	0.95	0.34
2nd quarter/1992		1.50	--	0.55	0.80	0.27
3rd quarter/1992		1.70	--	0.60	0.88	0.23
4th quarter/1992		0.30	--	0.01	0.01	0.10
1st quarter/1993		0.08	--	0.05	0.05	0.07

[†]Ponds were sampled weekly April 1989 to December 1990 and monthly from December 1990 to June 1991, and quarterly thereafter (except for the first and second quarters of 1992, where the values given are an average of three monthly determinations).

[‡]No discharge.

[§]na = not analyzed.

Proposed Guidelines for the Use of Biosolids on Surface Mine Lands

The results of this project indicate that the potential benefits of biosolids utilization on surface-mined lands far outweigh any environmental risks. In fact, we have observed no significant detrimental impact on soils, vegetation, or water quality in this study to date. While the long-term effects of biosolids utilization at this site will need continued monitoring, we feel that it is appropriate to move forward with this beneficial practice at additional sites in the Central Appalachian coal fields.

The following guidelines for the use of sewage sludge biosolids in reclamation are based on the results of this project, our parallel research work at the Powell River Project area since 1982, and our interpretations from relevant literature.

We recommend a one-time application of stabilized sewage sludge that has undergone appropriate pathogen reduction processes and meets the current Virginia–USEPA metal loading limits for the rate applied. If the material is to be applied as sewage sludge cake, the application rate should be $\leq$80 Mg ha^{-1} (35 ton acre^{-1}). If it is a composted or combined wood waste–sludge product with a C/N ratio of >20:1, then it should be applied at a rate $\leq$112 Mg ha^{-1} (50 ton acre^{-1}). Before the biosolids are applied, the area should mapped using an approach similar to our established mined land mapping system. Soils of the area should be sampled on a 0.5 ha grid for measurement of pH, extractable nutrients, and potential acidity (if necessary). Areas with a pH of <5.5 should be limed based on their measured potential acidity (acid-base account). Water on site and at discharge points must be monitored for nitrate and heavy metal levels for several months before application in order to establish baseline levels. During biosolids application, care must be taken to avoid spreading biosolids within 5 m of drainageways, outslopes, and wetland areas. Biosolids should be immediately incorporated into the soil to a depth of 15 cm or more. Nitrate and metal levels in ground water should be monitored monthly for a year or more after final spreading is complete, and nitrate should be monitored quarterly throughout the bond release period. Vegetation growing on site should be tested for metal content during the first growing season, and at the end of the 5-yr bond release period.

CONCLUSIONS

Results from the first 5 yr of this project indicate that the benefits of using municipal sewage sludge biosolids to reclaim mined lands far outweigh any environmental risks if biosolids are applied correctly and the soils, vegetation, and water in and around the site are carefully monitored after application. Biosolids are a superior treatment for mined land reclamation as long as the quality of the sludge is assured, and appropriate site selection, mapping, and monitoring techniques are utilized. The technical and economic issues involved with sewage sludge utilization on mined lands are relatively straightforward and easy to assess. Public acceptance is a major challenge in many communities, however, particularly when material from a distant source is to be utilized. Any large scale program for biosolids utilization on mined lands in the Appalachians will need to be accompanied by an effective public relations and education program. Large demonstration studies such as the one that we conducted at the Powell River Project are an important component of such a program, but a considerable amount of one-on-one interaction with local leaders and citizens will also be required for any program to be successful.

Additional research is needed in several areas. First of all, the optimum blending ratio of alternate waste C sources such as sawdust, bark mulch, newspaper waste, and papermill sludge has not been evaluated to date. As shown by this project, the inclusion of high C wood chips with the sludge cake can allow higher one-time reclamation loading rates, but the parameters that should be used to specify the C addition ratios for most types of substrates are not well defined. While our work has shown essentially no net movement of nitrates to local ground and surface waters, significant total-N losses from the surface soils were noted. Additional work on amended soil N balances and loss pathways is needed to verify that nitrate losses to the subsoil are indeed minimal and that our data do not in fact reflect large scale dilution factors at the treated sites.

ACKNOWLEDGMENTS

This project was funded by Enviro-Gro Technologies Inc., and the Virginia Center for Innovative Technology. We would also like to express appreciation for the support and cooperation of Penn-Virginia Resources, the Blackwood Land Company, Red River Coal Company, the Norfolk Southern Corporation, the Virginia Division of Mined Land Reclamation, the Virginia Department of Health, the Virginia Water Control Board, the Philadelphia Water Department and the Powell River Project. Thanks to Jay Bell and Jay Roberts for their work on the field experiment, and to Environmental Monitoring for water analyses.

REFERENCES

Daniels, W.L., and D.F. Amos. 1981a. Mapping, characterization and genesis of mine soils on a reclamation research area in Wise County, Virginia. p. 261–275. *In* Proc. 1981 Symposium on Surface Mining Hydrology, Sedimentology and Reclamation. Lexington, KY. 7-11 Dec. 1981. Univ. of Kentucky, Lexington.

Daniels, W.L., and D.F. Amos. 1981b. Minesoil maps and legend and mapping unit interpretations of the Powell River project reclamation research project. Dep. of Agronomy, Virginia Polytechnic Institute and State University, Blacksburg.

Daniels, W.L., and D.F. Amos. 1982. Chemical characteristics of some SW Virginia minesoils. p. 377–381. *In* Proc. 1982 Symposium on Surface Mining Hydrology, Sedimentology and Reclamation. Lexington, KY. 5-10 Dec. 1982. Univ. of Kentucky, Lexington.

Haering, K.C., and W.L. Daniels. 1991. Final report: Development of new technologies for the utilization of municipal sewage sludge on surface mined lands. Virginia Tech Res. Div., Blacksburg.

Haering, K.C., and W.L. Daniels. 1992. Development of new technologies for the utilization of municipal sewage sludge on surface mined lands. p. 83–89. *In* Proc. 1992 Powell River Project Symp., Wise, VA. 13 Sept. 1992. Virginia Tech Res. Div., Blacksburg, VA.

Moss, S.A., J.A. Burger, and W.L. Daniels. 1989. Pitch x loblolly pine growth in organically amended mine soils. J. Environ. Qual. 18:110–115.

Peterson, J.R., C. Lue-Hing, J. Gschwind, R.I. Pietz, and D.R. Zenz. 1982. Metropolitan Chicago's Fulton County sludge utilization program. p. 322–338. *In* W.E. Sopper et al. (ed.) Land reclamation and biomass production with municipal wastewater and sludge. Pennsylvania State Univ. Press, University Park.

Roberts, J.A., W.L. Daniels, J.C. Bell, and J.A. Burger. 1988a. Early stages of mine soil genesis as affected by topsoiling and organic amendments. Soil Sci. Soc. Am. J. 52:730–738.

Roberts, J.A., W.L. Daniels, J.C. Bell, and J.A. Burger. 1988b. Early stages of mine soil genesis in a SW Virginia mine spoil lithosequence. Soil Sci. Soc. Am. J. 52:716–723.

Roberts, J.A., W.L. Daniels, J.C. Bell, and D.C. Martens. 1988c. Tall fescue production and nutrient status on southwest Virginia mine soils. J. Environ. Qual. 17:55–62.

Sopper, W.E. 1992. Reclamation of mined land using municipal sludge. Adv. Soil Sci. 17:351–431.

Sopper, W.E., and E.M. Seaker. 1983. A guide for revegetation of mined land in the eastern United States using municipal sludge. Inst. for Land and Water Resour., Pennsylvania State Univ., University Park.

SECTION IV

CASE STUDY: THE ROSEMOUNT SEWAGE SLUDGE WATERSHED

17 Rosemount Watershed Study on Land Application of Municipal Sewage Sludge

W. E. Larson

Department of Soil Science
University of Minnesota
St. Paul, Minnesota

C. E. Clapp
R. H. Dowdy
D. R. Linden

USDA-Agricultural Research Service
University of Minnesota
St. Paul, Minnesota

Research on land utilization of sewage sludge was started in Minnesota in 1971 on farm fields with the objective of determining the nutrient availability and crop yield response to liquid-digested municipal sewage sludge. It was obvious from the start that the nitrogen (N) and phosphorus (P) in sludge were readily available to plants and that excellent yields could be obtained. After 2 yr of small plot work, it was thought that a holistic approach to sludge management on a field scale was needed.

A site was chosen at the Rosemount Experiment Station of the University of Minnesota. The site was available indefinitely, and was relatively isolated from homes and industry, but was close to the Minneapolis–St. Paul metropolitan area. The landscape with rolling topography, suitable for developing a closed watershed, offered a challenge in terms of controlling runoff and sediment movement and possible contaminants contained therein. The soils were developed from permeable loess over compacted glacial till.

The watershed study was initiated in 1973 with the primary goal to increase our knowledge about effects of liquid sewage sludge on surface and ground water quality, crop yields and quality, and soil quality over an extended period of time. Because the 16 ha (40 acre) site contained sloping land, soil and water control facilitates were constructed to contain all surface water within the watershed, and to minimize erosion. Sludge was applied at rates to meet the N needs of high producing corn (*Zea mays* L.) and reed canarygrass (*Phalaris arundinacea* L.). Control treatments receiving commercial fertilizer without sludge were also included (Knuteson et al., 1988; Halbach et al., 1994).

Sewage Sludge: Land Utilization and the Environment. SSSA Miscellaneous Publication.

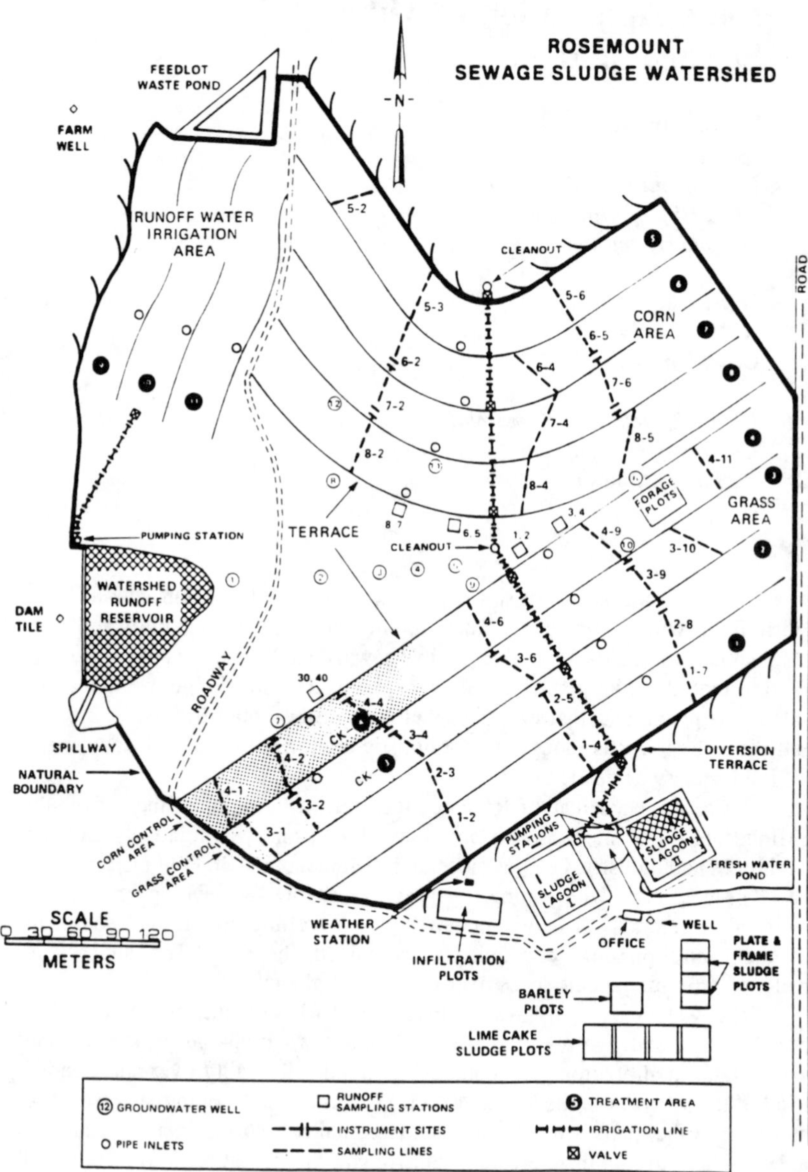

Fig. 17-1. General plan showing engineering design and sample collection locations for the Rosemount sewage sludge watershed.

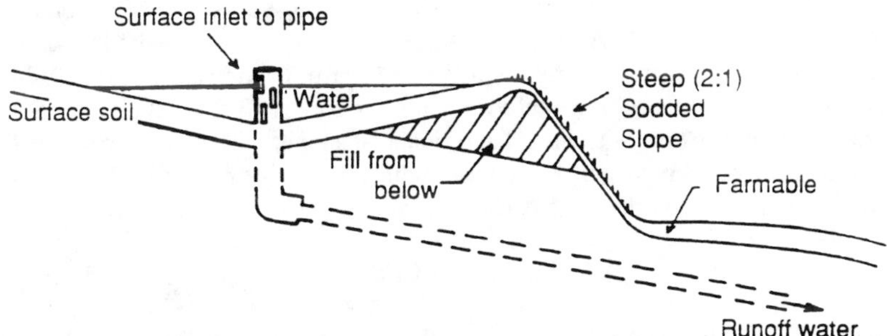

Fig. 17-2. Backslope terrace cross-section.

To control surface movement of soil and water, parallel graded terraces were constructed prior to sludge application (Fig. 17-1). The terraces were constructed with surface inlets that directed the water through underground pipe (tile) to a grassed waterway at the lower portion of the watershed (Fig. 17-2). Facilities were installed for sampling the runoff water from each pipe inlet. The equipment took samples of the water (and sediment) for analysis at intervals throughout a runoff event. A water reservoir at the outlet to the watershed contained all runoff water within the study area.

The study site is in an area of complex glacial moraine deposits. The gradient of slopes in the watershed range from 1 to 10%. Soils on the site were formed in wind-deposited loess, which overlays dense glacial till. Two sludge storage lagoons were constructed adjacent to the watershed so that liquid sludge delivered by tank truck from several suburban plants could be stored at the site until time of application. At one or two times each year (fall or spring) sludge was injected into the soil on land used for corn production. Sludge was applied to the surface on the reed canarygrass areas. Initially the sludge was applied to the grassed areas by tank wagon; later by a traveling irrigation gun. It was applied when the grass was dormant and immediately after harvest during the growing season. The area received potassium (K) fertilizer as needed as evidenced by soil tests.

The area was farmed with conventional practices to produce high yielding crops. Reed canarygrass yields averaged near 11 Mg ha^{-1} (4.9 ton acre^{-1}) and corn grain 8.4 Mg ha^{-1} (156 bu acre^{-1}). Grass was harvested for hay three times per year and corn (fodder) for ensilage.

An extensive scheme for sampling water, soil, and plant materials was used. In addition to sampling the tile drainage from the terraces, porous ceramic samplers for extracting soil water were located strategically throughout the study area. Shallow wells for sampling free water at the contact between the loess and compact glacial till were also used. In addition water from two nearby deep wells were sampled. Transects across the terrace intervals for sampling soils and vegetation have been used and maintained throughout the life of the study. Details of the sampling procedures, laboratory analysis, and results are given in accompanying chapters.

The study has featured a close inter-agency cooperation. The University of Minnesota; the USDA–Agricultural Research Service, and the USDA–Soil Conservation Service; the Metropolitan Waste Control Commission of the Twin Cities; and the U.S. Environmental Protection Agency (USEPA) all contributed substantially in the form of technical and financial support. The University-ARS research team was given USEPA's outstanding research award for Beneficial Sewage Sludge Use on the Watershed in 1989.

REFERENCES

Halbach, T.R., C.E. Clapp, R.H. Dowdy, W.E. Larson, and S.R. Thomas. 1994. Agricultural utilization of sewage sludge: A 20-year study at the Rosemount Agricultural Experiment Station. Minnesota Agric. Exp. Stn. Misc. Publ. St. Paul.

Knuteson, J., C.E. Clapp, R.H. Dowdy, and W.E. Larson. 1988. Sewage sludge management. Land application of municipal sewage sludge to a terraced watershed in Minnesota. Minnesota Agric. Exp. Stn. Misc. Publ. 56–1988. St. Paul.

18 Metropolitan Waste Control Commission's Perspective of the Rosemount Sewage Sludge Watershed

R. C. Polta

Metropolitan Waste Control Commission
St. Paul, Minnesota

In 1967, the Minnesota Legislature established the Metropolitan Council (MC) to serve as the seven county area's coordinating and development agency. In 1969, the Legislature adopted the Metropolitan Sewer Act, which directed the MC to adopt a comprehensive sewer plan for the metro area and to establish the Metropolitan Sewer Board (MSB) as an operating agency to implement the plan. By 1970 the MSB acquired 33 wastewater treatment plants, 320 miles of interceptor sewer, and other ancillary facilities.

Most of the 33 facilities had design flows of one million gallons per day (MGD) or less and only two had design flows of more than 10 MGD. The facilities that were inherited did thus not provide for cost effective treatment. More importantly, however, most of the facilities were not capable of meeting the existing secondary treatment standards. As a result, the MSB immediately initiated an aggressive program to consolidate and upgrade treatment facilities. The agency's name was changed to the Metropolitan Waste Control Commission (MWCC) in 1975; however, the goal of 100% compliance with environmental standards did not change. During 1992, the MWCC operated 11 wastewater treatment plants and obtained 99.9% compliance with all effluent standards for the year (two exceedances out of a total of 1904).

Although the initial efforts of the MSB staff was on upgrading the capabilities to treat wastewater, they realized that sludge disposal practices can have a significant impact on the environment. With the exception of the MSB's Metro Plant, which was incinerating sludge in 1970, all municipal wastewater treatment residuals generated in the State of Minnesota were disposed of on, or in, land. At that time the disposal of sludge on land was generally unregulated and it was not uncommon to find that sludges had been land applied inappropriately.

Sewage Sludge: Land Utilization and the Environment. SSSA Miscellaneous Publication.

In the early 1970s, the MSB staff realized that land application might be a viable long term sludge disposal–utilization alternative; however, at that time the impacts of sludge utilization on agricultural land had not been documented. Although sludge had been spread on farmland for many years in Minnesota, as well as other northern states, the data on impacts to crops and the soil and water environment was largely limited to anecdotal information. They also realized that land application would soon be regulated by state and federal agencies as well as by public opinion. The MSB thus initiated joint projects with state and federal researchers to determine if land application of sludge could be accomplished in Minnesota without detrimental impacts on crops and the soil and water environments.

Starting in 1971, the MSB participated in cooperative efforts, with staff from the USDA-Agricultural Research Service and the University of Minnesota Soil Science Department, to investigate the fate of heavy metals and several other topics related to land application of sludge. In February of 1972, MSB Resolution 72-26 officially initiated participation in the Rosemount sludge project. Since that time the MWCC's direct support for the project totals approximately $1.4 million.

The participation of the MSB was based on several broad goals. The primary goals were to identify and document the agricultural and environmental impacts of sludge application to crop lands, in particular nutrient availability and utilization rates and the fate of metals (crop, soil, ground water, or surface water). It was anticipated that sludge could be used on agricultural land without generating environmental or aesthetic problems. One of the secondary goals was thus to publicly demonstrate the results of sludge utilization in the hopes that, at some point in the future, there would be sufficient interest in the local agricultural community to support land application as a potentially viable alternative to incineration. The last goal was to develop expertise in the techniques and processes necessary to utilize sludge on crop land. It was anticipated that a number of graduate students would work on the project and that some of these students would eventually work directly for the MSB/MWCC as well as local (county and state) or federal regulatory agencies. It was believed that these knowledgeable staff would benefit both the regulatory and operating agencies.

In 1978 the MWCC initiated an interim land application program because of limited incinerator capacity at the Metro Plant. At that time, the probability for successfully operating a large scale land application program in the metropolitan area was not considered to be very high. Although the program started small, in 1981 ≈200 000 wet tons of dewatered sludge was applied to area farmland. The ability for county agents and other technical staff, as well as local officials, to visit the Rosemount site and discuss their concerns with the knowledgeable project staff was a significant factor in terms of gaining public acceptance of land application in Minnesota at that time. It is unlikely that the MWCC's program would have been successful without the Rosemount project. The interim land application program was terminated in 1984 when the new incinerators at Metro were all put into service.

In 1981 the Minnesota Pollution Control Agency adopted formal sludge management rules. These rules provided additional credibility for the MWCC's land application program because they served as standards against which people could compare the MWCC's performance. It must be noted that the data generated at Rosemount served as the basis for much of the rules.

The MWCC's initial goals have all been obtained. The topics of nutrient utilization and the fate of metals are discussed in subsequent chapters in this book by Clapp et al. (1994) and Dowdy et al. (1994), respectively. As a result of the Rosemount demonstration and the subsequent interim program, land application is now routinely considered to be an alternative to incineration. Finally, junior project staff eventually took positions with regulatory agencies and the MWCC. Their experience at Rosemount, no doubt, served them well in terms of both regulating and operating sludge utilization facilities.

REFERENCES

Clapp, C.E., R.H. Dowdy, D.R. Linden, W.E. Larson, C.M. Hormann, K.E. Smith, T.R. Halbach, H.H. Cheng, and R.C. Polta. 1994. Crop yields, nutrient uptake, soil and water quality during 20 years on the Rosemount sewage sludge watershed. p. 137–147. *In* C.E. Clapp et al. (ed.) Sewage sludge: Land utilization and the environment. SSSA Misc. Publ. SSSA, Madison, WI (this publication).

Dowdy, R.H., C.E. Clapp, D.R. Linden, W.E. Larson, T.R. Halbach, and R.C. Polta. 1994. Twenty years of trace metal partitioning on the Rosemount sewage sludge watershed. p. 149–155. *In* C.E. Clapp et al. (ed.) Sewage sludge: Land utilization and the environment. SSSA Misc. Publ. SSSA, Madison, WI (this publication).

19 Design of Conservation Structures at the Rosemount Sewage Sludge Watershed

J. A. Jeffery

USDA-Soil Conservation Service
Brooklyn Center, Minnesota

Due to topography, soils, crops, and climatic conditions, utilization of municipal sludge on agricultural land will often require structural conservation practices. These practices are needed for water and sludge management, to control erosion, to store sludge for application, and to collect and retain all runoff from the application area.

A watershed of ≈16 ha (40 acres) at the Rosemount Agricultural Experiment Station of the University of Minnesota was selected to study the application of sewage sludge to land. Moderately complex slopes in the watershed range from 0 to 12%. The soils of the watershed have from 60 to 240 cm (24 to 90 in) of a silt loam loess overlying compact glacial till. Port Byron, (fine–silty, mixed, mesic Typic Hapludoll), Bold [coarse–silty, mixed (calcareous), mesic Typic Udorthent], and Tallula (coarse–silty, mixed, mesic Typic Hapludoll) are the dominant soil types. A 200-head beef (*Bos* sp) feedlot was near one corner of the watershed.

The general design was based on a visual inspection of the watershed that indicated the areas to be used for research plots, sludge storage ponds, runoff reservoir, runoff disposal, and runoff sampling stations. The following design criteria were established to meet the objectives of the study:

1. Use parallel terraces constructed with a minimum of land forming.
2. Provide a detention reservoir to collect and hold the runoff from all the area that could come in contact with sludge applications either directly or indirectly.
3. Provide a disposal area for the collected runoff.
4. Provide sludge storage sufficient for an application of ≈7.5 cm (3 in) in the spring before planting.
5. Divert runoff from all nonresearch areas of the watershed away from the sludge-treated area.
6. Provide an underground outlet to direct the collected runoff from each terrace to a sampling station and then through a drainageway to the

Sewage Sludge: Land Utilization and the Environment. SSSA Miscellaneous Publication.

runoff reservoir.

7. Design the site to contain a 100-yr, 24-h storm runoff of 112 mm (4.4 in).

The USDA-Soil Conservation Service, in cooperation with the USDA-Agricultural Research Service, the Metropolitan Waste Control Commission, Minnesota Agricultural Experiment Station, and the Minnesota Pollution Control Agency, then developed a research area design based on the general layout plan and the above design criteria.

DESIGN

Terraces

Steep grassed backslope storage terraces with underground outlets were used. Terraces were spaced 39.6 m (130 ft) apart and were designed to impound 64 mm (2.5 in) of runoff.

Terrace channels were graded to vertical plastic intake pipes located at natural low points in the terrace channel. The protruding part of the intake pipe was perforated so that when runoff level reached design height it would allow a minimum discharge of 103 m^3 (1 acre-in) per day to prevent crop damage from standing water. Plastic underground outlet pipes discharged intake flow through a sampling station to shallow field ditches below the lowest terraces.

Graded diversions were installed above the terraces to divert natural runoff from the research area.

Runoff Reservoir

The runoff reservoir was formed by a compacted earth fill embankment across the watershed outlet. The embankment was keyed into the natural ground with a core trench extending the full length of the fill. The reservoir was designed to retain all runoff from the terraces and the service area below the terraces, a watershed of 20 ha (50 acres). The design reservoir capacity, prior to outflow, was to store the watershed runoff from a 25-yr, 24-h duration storm, 9.9 by 10^3 m^3 (8 acre-ft), and a 100-yr, 24-h duration storm, 22.6 by 10^3 m^3 (18.3 acre-ft). The reservoir was planned to be pumped empty whenever the stored runoff reached the 25-yr storm runoff volume. Therefore, there would always be storage capacity to contain the entire 100-yr storm runoff without outflow. The bottom of the reservoir was lined with a black 0.2–mm polyethylene film to the 25-yr storage level.

A vegetated emergency spillway was constructed to provide for outflow and to maintain embankment integrity should the design runoff be exceeded.

Sludge Storage Ponds

Two separate sludge storage ponds were formed by a combination of excavation and compacted earth fill dikes. The storage ponds were 2 m (6.5 ft) deep, with 55 m (180 ft) square bottoms and 3:1 slopes on the inside and 2:1 slopes on the outside. The combined capacity of the ponds was 11 400 m^3 (3 000 000 gal), the approximate winter storage capacity needed for sludge expected from four treatment plants supplying sludge for the study. About 0.460 m (18 in) additional storage was provided in each pond for residual solids that could not be pumped and contingent sludge volumes. The pond bottoms were level except for a 6 m (20 ft) square sump 0.6 m (2 ft) deep at the corner adjacent to the pumping stations. One pond was lined with a black 0.2–mm polyethylene film; the other was unlined. This was to provide a comparison to determine whether the solids seal the bottom of ponds sufficiently to prevent seepage and eliminate the need for such lining.

Feedlot Waste Storage Pond

A waste storage pond was installed below the beef feedlot area to isolate the livestock waste from the research area. A storage diversion was constructed above the waste storage pond to divert all runoff from the livestock area of the watershed into the pond. All runoff collected in this waste storage pond was planned to be pumped to a pasture outside the research watershed.

CONSTRUCTION

The Metropolitan Waste Control Commission performed construction contract administration and the Soil Conservation Service provided construction layout and inspection services. Bids were received resulting in Metro Engineering, Osseo, MN, being the successful bidder. Installation was completed in Fall 1973 for $79,000.

EVALUATION

Terraces

The terraces have controlled runoff, channeling it to the intake as planned. No serious erosion problems have occurred. A high percentage of transported sediment (soil and sludge) settled in the terrace storage area and did not enter the underground outlet.

Occasionally, the storage area around the vertical inlet remained wet for a few days after larger runoff events. Although these areas interfered temporarily with farming operations and depressed crop growth around the inlet, they are insignificant relative to the total crop area. A drain tile in the storage area channel connected to the underground outlet at the vertical inlet would reduce or eliminate this wetness condition.

The curved terraces required additional equipment and labor to make the application units follow curved sections. The traveling sprinkler and subsurface injector are best suited to straight line operation. Straight terraces or water and sediment basins need to be used whenever possible. Where the traveling sprinkler applying sludge had to travel downslope in the terrace channel, application rates, satisfactory for other parts of the field, overtook the sprinkler creating wheel furrows. This situation can be solved by providing a travel path out of the channel or closely monitoring application rates in this area.

Observations indicated that at the start of each growing season it was advisable to remove any accumulated sediment from the terrace storage area.

Runoff Reservoir

The runoff reservoir capacity was sufficient to contain all runoff events from the watershed. The emergency spillway has not flowed.

The reservoir provided a source of irrigation water during dry periods. It also served as a water source for cleaning the sludge application system.

Sludge Storage Ponds

Samples at various depths under and outside the ponds indicated no seepage. The pond bottoms were apparently effectively sealed. In this case, the polyethylene film liner may not have been needed.

In a small area on the pond side slope, the 25 cm (1 ft) of soil protection over the polyethylene liner slid into the pond. This indicated that the side slope may need to be flatter than 3:1.

SUMMARY

A complete structural conservation system applied on the watershed effectively provided surface water management, controlled erosion, provided adequate sludge storage between application periods, and retained the runoff from all storm events that occurred on the application area. Practices applied were those generally accepted by the agricultural community for water and sediment control on cropland.

The terraces controlled the runoff and, with the perforated vertical inlet, allowed the soil and sludge solids to satisfactorily settle out of the runoff before it was discharged. Wet spots in the crop area were minimal and can be reduced or eliminated by installing drain tile. The terraces caused only minor isolated interference with normal agronomic practices. Terrace spacing significantly reduced erosion on the sloping cropland.

The curved terraces caused some difficulty in making application with a mobile hose in conjunction with the traveling sprinkler or subsurface injector. Techniques for use of this type of application need further development. Although sludge storage pond lining may not have been necessary at this location, lining to control seepage may be critical in other applications. Further study is needed to improve existing design aids.

20 Crop Yields, Nutrient Uptake, Soil and Water Quality During 20 Years on the Rosemount Sewage Sludge Watershed

C. E. Clapp
R. H. Dowdy
D. R. Linden

USDA-Agricultural Research Service
University of Minnesota
St. Paul, Minnesota

W. E. Larson
C. M. Hormann
K. E. Smith
T. R. Halbach
H. H. Cheng

Department of Soil Science
University of Minnesota
St. Paul, Minnesota

R. C. Polta

Metropolitan Waste Control Commission
St. Paul, Minnesota

Disposal of the products of municipal wastewater treatment is a major environmental problem. Finding environmentally acceptable, socially responsible, and economically feasible plans for carrying out this task is receiving much attention from both research and regulatory agencies, as well as the public. Increasingly rigid water and air quality standards has forced municipal officials, environmental engineers, and wastewater treatment plant operators to look beyond conventional systems of treatment and disposal. Land application of sewage sludge and effluent is an alternative that returns materials to a natural cycle, which is agriculturally beneficial. In turn, renewed interest in treatment on land has stimulated scientists to examine in detail the physical, chemical, and biological processes in soil that influence waste renovation. Historically, land treatment methods have been based on sewage sludge disposal, whereas we now emphasize a more modern approach to wastewater treatment followed by recycling on agricultural land.

Sewage Sludge: Land Utilization and the Environment. SSSA Miscellaneous Publication.

The Rosemount watershed study was initiated in 1973 on land of the University of Minnesota's Agricultural Experiment Station near Rosemount, MN (Clapp et al., 1977, 1983). The primary goal of the research was to develop efficient, practical, and environmentally safe methods for utilizing municipal sewage sludge on land in harmony with agricultural usage. The long-term study at the Rosemount watershed is one of the best examples in the world of a detailed analysis of environmental and agricultural impacts from sludge application to land. After 20 yr of research, results have shown that there are many benefits from using sludge as a plant nutrient source for corn (*Zea mays* L.) and grass crops. Historically, yields on the sludge-applied land have been slightly better than those on the fertilized control areas within the same watershed (Knuteson et al., 1988).

MATERIALS AND METHODS

A 16-ha watershed was terraced to give 10 treatment areas with separate tile inlets for surface runoff water. (See Fig. 17-1 of Larson et al., 1994, this publication.) Four sludge areas and a control area were planted to corn; four more sludge areas and a control area were planted to reed canarygrass (*Phalaris arundinacea* L.). Digested municipal sewage sludge was applied annually and details of storage, application methods, and soil and crop management are given by Duncomb et al. (1982) and Knuteson et al. (1988).

Soils of the watershed are 60 to 240 cm of silt loam loess overlaying compact glacial till. Port Byron (fine-silty, mixed, mesic Typic Hapludoll), Bold [coarse-silty, mixed (calcareous) mesic Typic Udorthent], and Tallula (coarse-silty, mixed, mesic Typic Hapludoll) are the dominant soil series. Total carbon (C), total nitrogen (N) and pH of the surface soils, prior to the initiation of the study in 1973, ranged from 1.5 to 4.0%, 0.12 to 0.34%, and 5.3 to 7.5, respectively. The cation–exchange capacities of the soils ranged from 20 to 25 $cmol_c$ kg^{-1}.

Corn was harvested along three sampling lines of each corn area (15 plots). Grain was picked and the rest of the plant was chopped as stover. Reed canarygrass samples were harvested at three cutting times: in early June, mid-July, and early September. Plots were located at three sites (in the terrace channel, at the midterrace interval, and near the upper terrace interval) along three sampling lines of each grass area (45 plots).

Composite soil samples at 0- to 15- and 15- to 30-cm depths were taken in the fall of 1973 after the watershed was constructed, along sampling lines designated on each terrace interval. The same sites were resampled in 1978, 1983, 1988, and 1993. Annual fall samples were regularly taken at the 0- to 15-cm depth for soil test analysis.

The terrace channels had surface tile inlets that directed runoff water underground to an outlet at a central drainage channel. Runoff water on each area was collected by automatic samplers modified to begin collection when flow started and at 1-h intervals during runoff. Flow rates were measured by a water stage recorder with slotted tube and stilling well (Linden et al., 1983). Soil water was sampled at 3- to 4-wk intervals by porous ceramic samplers installed at 60- and 150-cm depths at 24 sites on the sludge and control areas. Samples from 14 shallow ground water monitoring wells within the watershed were collected monthly. Background samples from various water sources, both within and around the watershed, were taken at regular intervals for the 1973 season before any sludge was applied on the project site, and compared with water samples taken during the following growing seasons.

Water and sludge samples were refrigerated immediately after collection, and analyzed for organic components within 1 wk or the samples were acidified. Soil samples were oven dried at 35°C; plant samples were dried at 65°C. Sludge, plant, soil, and water analytical methods are described in Knuteson et al. (1988).

RESULTS AND DISCUSSION

During the 20-yr study, municipal sewage sludge was applied on corn areas at an average rate of 4 cm liquid yr^{-1}, 11 Mg solids ha^{-1} yr^{-1}, and 475 kg N ha^{-1} yr^{-1}. Total of 28 applications was 68 cm of sludge, 224 Mg ha^{-1} solids, and 9500 kg ha^{-1} total N. On grass areas for 12 yr, sludge was applied at an average rate of 8 cm liquid yr^{-1}, 15 Mg solids ha^{-1} yr^{-1}, and 830 kg N ha^{-1} yr^{-1}. Total of 61 applications was 96 cm of sludge, 175 Mg ha^{-1} solids, and 10 000 kg ha^{-1} total N. Total sludge and nutrients applied by cropping season are shown in Table 20–1. Sludge analyses are given in Table 20–2.

Table 20-1. Sludge and nutrients applied by cropping season for Rosemount sewage sludge watershed (1974–93).

Season	Applica-tions	Sludge	Solids		Nutrients		
					N	P	K
		cm	%	Mg ha^{-1}	---------kg ha^{-1}-----------		
			Corn				
1974-77	11	20.3	1.40	39.1	2220	1180	190
1978-81	7	24.0	2.75	67.1	3450	1680	191
1982-85	3	10.8	2.08	30.3	1440	671	62
1986-89	3	6.6	2.25	23.4	936	607	72
1990-93	4	6.7	9.50	64.0	1410	973	89
Total	28	68.4	----	224	9460	5100	604
			Grass†				
1974-77	18	24.3	1.76	43.2	2530	1110	254
1978-81	33	50.0	1.95	101	5570	2420	370
1982-85	10	21.9	1.31	28.6	1940	600	79
Total	61	96.2	----	173	10040	4130	703

†Grass area planted to corn since 1986.

Table 20-2. Composition of sludge applied to corn and grass treatment areas for Rosemount sewage sludge watershed (1974–1992).

Constituent†	Corn		Grass
	1974–1989	1990–1992	1974–1985
	%		
Total solids	2.96	10.0	1.94
Total C	23.6	15.0	25.7
Total N	6.92	1.83	5.93
NH_4–N	1.77	0.63	2.58
P	2.51	1.75	2.46
K	0.33	0.17	0.43
	$dS\ m^{-1}$		
Electrical conductivity	4.0	3.3	4.4
pH	7.5	7.5	7.8

†Total solids based on 105°C dry weight. Other constituents based on percentage of total solids.

Liquid sewage sludge was an effective source of N and P for corn and forage crops. Yields equalled or surpassed those of areas with traditional fertilizer management (Fig. 20-1). Twenty-year average corn grain and fodder yields were 8.4 Mg ha^{-1} (156 bu $acre^{-1}$) and 17.4 Mg ha^{-1}, and 7.7 Mg ha^{-1} (145 bu $acre^{-1}$) and 15.9 Mg ha^{-1} for the sludge and control areas, respectively (Table 20-3). Average annual reed canarygrass yields for 12 yr were 11.0 and 9.6 Mg ha^{-1} for the sludge and control areas, respectively (Table 20-4).

Elemental analyses of plant tissue showed that corn took up N, P, and K equally well from sludge- and control-managed areas, removing an average of 205 and 184 kg N ha^{-1} yr^{-1} for the sludge and control areas, respectively. Comparison of N applied and N removed for corn sludge and control areas for 1974 to 1992 is made in Fig. 20-2. The reed canarygrass removed 335 and 286 kg N ha^{-1} yr^{-1} for the same respective treatments. The grass tissue showed less-than-optimum amounts of P available in the grass control area.

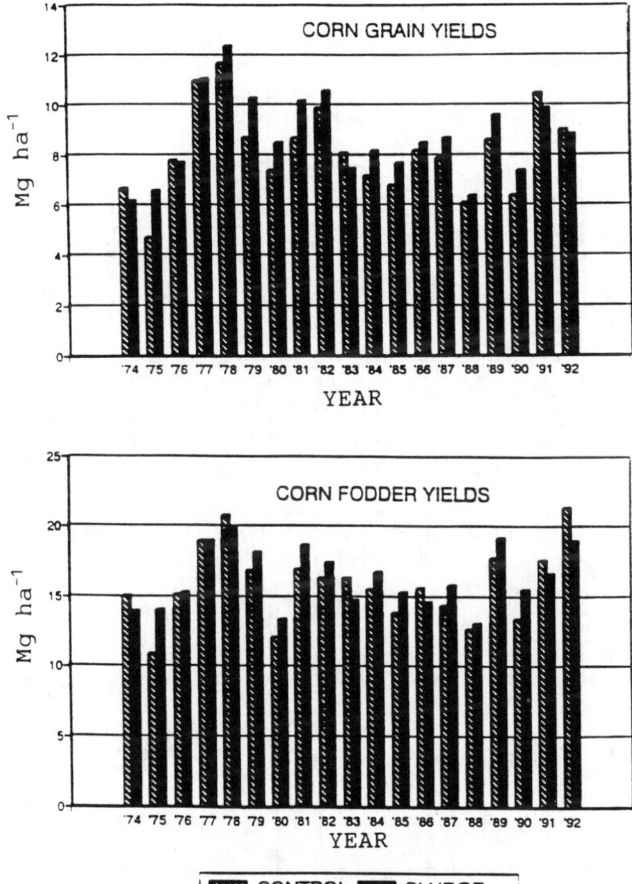

Fig. 20-1. Corn grain and fodder yields for control and sludge areas (1974–1992).

An important point to note is that large amounts of N and P applied by sludge did not adversely affect yields by creating imbalances of nutrients within the plant. Sludge also supplied many of the other essential plant nutrients. Because N and P removed from sludge- and control-treated areas were nearly equal for corn, it appeared that lower sludge application rates could be used, thus increasing the efficiency of nutrient use by the crop.

Table 20-3. Corn yields and nutrients removed for Rosemount sewage sludge watershed (1976–1992).

Treatment	Year	Yield†			Total nutrients removed		
		------ Grain -----		Fodder	N	P	K
		$Mg\ ha^{-1}$	$bu\ acre^{-1}$	$Mg\ ha^{-1}$	----- $kg\ ha^{-1}$ -----		
Control	1976–1979	8.3	156	18.0	203	32	157
	1980–1983	7.6	142	15.5	195	32	142
	1984–1987	7.4	139	14.7	169	23	120
	1988–1992	7.9	149	16.6	187	28	105
Mean		7.7	145	15.9	188	29	130
Sludge	1976–1979	8.8	165	18.1	223	38	165
	1980–1983	8.1	152	19.4	232	38	152
	1984–1987	8.3	155	15.7	187	36	127
	1988–1992	8.3	156	16.7	183	38	104
Mean		8.4	156	17.4	205	38	135

† Yields represent dry matter at 65°C; shelled grain calculated in $bu\ acre^{-1}$ at 15.5% H_2O.

Table 20-4. Reed canarygrass yields and nutrients removed for Rosemount sewage sludge watershed (1976–1983).

Treatment	Year	Dry matter yield				Total nutrients removed		
		Cut 1	2	3	Total	N	P	K
		---------- $Mg\ ha^{-1}$ ----------				--------- $kg\ ha^{-1}$ ---------		
Control	1976–1979	4.5	3.1	1.8	9.4	294	30	284
	1980–1983	5.3	2.6	1.8	9.7	279	33	305
Mean		4.9	2.8	1.8	9.6	286	32	294
Sludge	1976–1979	5.2	3.4	2.3	10.9	355	42	341
	1980–1983	5.6	3.5	2.2	11.3	328	52	390
Mean		5.4	3.4	2.2	11.1	342	47	366

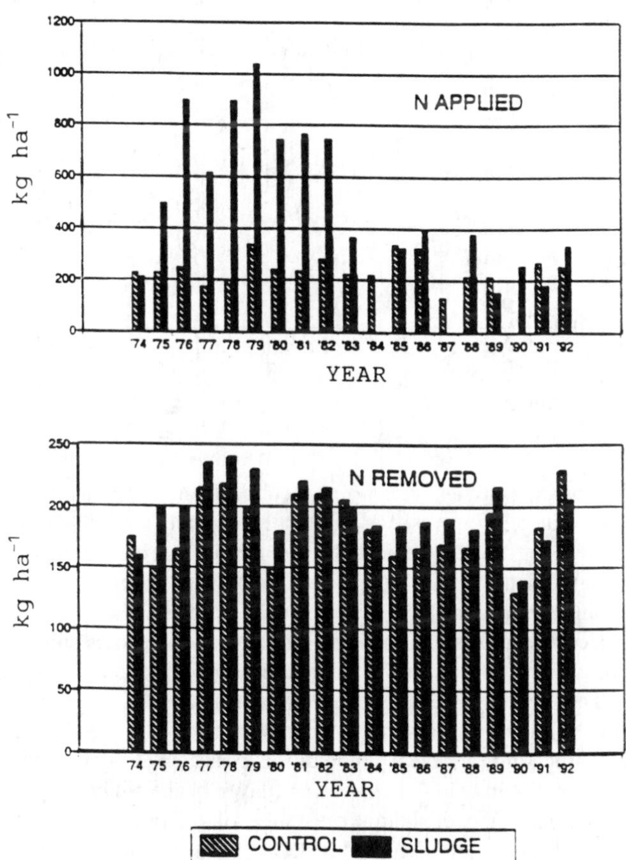

Fig. 20-2. Total N applied compared with N removed by corn on control and sludge areas (1974–1992).

Total soil C and N increased as expected on the sludge areas, with lower terrace positions having higher accumulations because of some downslope movement of sludge and soil. By 1993, total C, total N, and pH values on sludge-treated areas ranged from 1.9 to 4.8%, 0.14 to 0.41%, and 6.0 to 7.6, respectively. Control areas remained relatively unchanged for C and N during the 20-yr period, however soil pH decreased under control corn treatments, requiring lime application. Electrical conductivity varied where sludge was applied, but did not increase to levels adversely affecting plant growth. For comparative results, see Stark and Clapp (1980) and Harding et al. (1985).

Table 20-5. Summary of University of Minnesota soil test results of soil samples from Rosemount sewage sludge watershed (1992).

Characteristic	Units	Control Ck-4†	Sludge 1,2,3,4†	Sludge 5,6,7,8†
pH		6.2	6.6	6.8
Organic matter	%	3.3	4.7	3.8
Extractable P	kg ha^{-1}	100	>200	>200
Exchangeable K	kg ha^{-1}	360	260	250
Soluble salts	dS m^{-1}	0.3	0.3	0.2

†See diagram of watershed (Fig. 17-1 in Larson et al., 1994, this publication).

Composite soil samples were taken each fall from the sludge and control treatment areas after crop harvest for soil test analyses (Table 20-5). Soil pH decreases on the control areas indicated need for lime, which was applied in 1986. Soil organic matter on the sludge-treated areas increased over that of the control area. Sludge additions also increased extractable soil P levels to a range beyond that necessary for maximum crop production. The excess soil P did not affect crop production by reducing the availability of ions such as Zn, Cu, Ca, and Mg, since these cations were supplied by the sludge. Soil P on the control area was also very high due to commercial fertilizer additions. Exchangeable K was adequate, indicating sufficient amounts of inorganic K were supplied as fertilizer. Soluble salts in the 0- to 15-cm soil layer of the sludge-applied corn and grass areas remained well below the range of tolerance for field crops. Changes in soil properties are discussed in Clapp et al. (1986).

Protection of water quality is an important consideration when land applying fertilizers and sludge as soil amendments. Erosion, runoff, and leaching can be problems on sloping land used for row crops. Sediment and nutrients adsorbed on soil particles can cause environmental problems in surface waters. Controlling erosion, as was done on this Rosemount watershed, prevented excessive sediment loss in runoff waters. Incorporation of fertilizers and injection of sludge into the soil was also important in reducing loss of nutrients in runoff (Table 20-6). Crop type also had an effect on nutrient movement. Runoff waters from reed canarygrass treatment areas had higher P and K levels than those from corn areas, in spite of the fact that corn areas had much less surface protection during the winter (Dowdy et al., 1980). Trace element levels were very low in runoff from the watershed, except when sludge was surface-applied to grass in winter. Some trace metals leached out of the sludge mat on the frozen grass areas. Overall, surface water quality was very good at the study site.

Table 20-6. Analysis of runoff and snowmelt water and average annual losses for Rosemount sewage sludge watershed (1976–1981).

Treatment	Amount	Concentration		Loss	
		N	P	N	P
	cm	------ $mg\ L^{-1}$ -------		---$kg\ ha^{-1}\ yr^{-1}$ ---	
Corn runoff					
Control	8.6	3.9	0.4	3.3	0.3
Sludge	9.2	9.7	0.8	8.9	0.8
Grass runoff					
Control	6.0	6.8	1.1	4.1	0.7
Sludge	7.3	19.7	5.0	14.4	3.6
Grass snowmelt					
Control	3.2	8.0	0.8	2.6	0.2
Sludge	3.7	23.9	3.1	8.8	1.1

Table 20-7. Analysis of soil water for Rosemount sewage sludge watershed (1976–1981).

Treatment	Depth	Concentration	
		NO_3–N	PO_4–P
	cm	------ $mg\ L^{-1}$ ------	
Corn	60		
Control		116	0.03
Sludge		173	0.04
Grass	60		
Control		20	0.01
Sludge		90	0.13
Corn	150		
Control		97	0.02
Sludge		160	0.02
Grass	150		
Control		22	0.00
Sludge		52	0.05

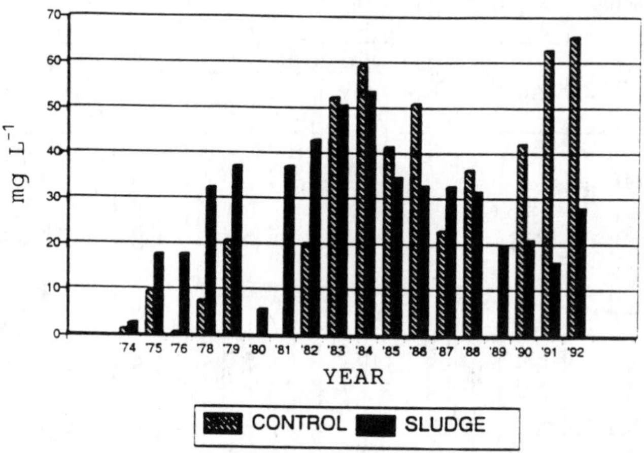

Fig. 20-3. Nitrate-N concentrations in water samples from shallow wells (1974–1992).

High levels of nitrates reduced quality of soil water (Table 20-7), but not deep ground water. Other water quality parameters were not affected by activities at the site. Excessive nitrates occurred due to applications of N which were in excess of crop needs in the early years (Fig. 20-3). Nitrate levels in the shallow ground water decreased in recent years in response to decreased application rates. Decreased application rates, however, did not reduce yields. At this site, nitrates did not enter deep ground water because of dense glacial till overlaying the deep aquifer (King et al., 1986). Periodic monitoring of the small stream below the watershed showed no degradation of water quality over the study period.

The long-term study of the Rosemount sewage sludge watershed is an excellent example of environmental and agronomic analysis of sludge application to land. The value of sludge as a nutrient source was established for corn and grass crops. Grass was more efficient than corn in removing N, P, and K supplied in the sludge. Choice of crop is important if land application of sludge is primarily a disposal method. From a water quality viewpoint, the Rosemount watershed study showed that sludge application can be conducted in an environmentally safe fashion. Surface water quality was protected by adequate soil erosion and runoff control. Monitoring shallow ground water allowed adjustments to management practices that protected the deep aquifer. In conclusion, this study is one of a very few to address both the agronomic and environmental issues concerning sludge application to land at the watershed scale over an extended period of time.

REFERENCES

Clapp, C.E, D.R. Duncomb, W.E. Larson, D.R. Linden, R.H. Dowdy, and R.E. Larson. 1977. Crop yields and water quality after application of sewage sludge to an agricultural watershed. p. 185–198. *In* R.C. Loehr (ed.) Food, fertilizer, and agricultural residues. 9th Annual Cornell Agric. Waste Management Conf., Syracuse, NY. 27–29 Apr. 1977. Ann Arbor Sci. Pub., Ann Arbor, MI.

Clapp, C.E., W.E. Larson, R.H. Dowdy, D.R. Linden, G.C. Marten, and D.R. Duncomb. 1983. Utilization of municipal sewage sludge and wastewater effluent on agricultural land in Minnesota. p. 259–292. *In* K. Schallinger (ed.) Proc. 2nd Int. symp. on peat and organic matter in agriculture and horticulture. Bet Dagan, Israel. 9–14 Oct. 1983. Inst. Soil and Water, Volcani Center, Bet Dagan, Israel.

Clapp, C.E., S.A. Stark, D.E. Clay, and W.E. Larson. 1986. Sewage sludge organic matter and soil properties. p. 209–253. *In* Y. Chen and Y. Avnimelech (ed.) The role of organic matter in modern agriculture. Martinus Nijhoff Publ., Dordrecht, The Netherlands.

Dowdy, R.H., C.E. Clapp, D.R. Duncomb, and W.E. Larson. 1980. Water quality of snowmelt runoff from sloping land receiving annual sewage applications. p. 11–15. *In* Proc. of national conf. on municipal and industrial sludge utilization and disposal. Alexandria, VA. 28–30 May 1980. Information Transfer Inc., Silver Spring, MD.

Duncomb, D.R., W.E. Larson, C.E. Clapp, R.H. Dowdy, D.R. Linden, and W.K. Johnson. 1982. Effect of liquid wastewater sludge application on crop yield and water quality. J. Water Pollut. Control Fed. 54:1185–1193.

Harding, S.A., C.E. Clapp, and W.E. Larson. 1985. Nitrogen availability and uptake from field soils five years after addition of sewage sludge. J. Environ. Qual. 14:95–100.

King, T.A., J.B. Swan, C.E. Clapp, R.H. Dowdy, W.E. Larson, and R.C. Polta. 1986. The long-term nutrient budget of a sewage sludge lagoon. p. 204–216. *In* Proc. 9th Annual Madison Conf.on Municipal and Industrial Waste, Madison, WI. 9–10 Sept. 1986. Univ. of Wisconsin, Madison, WI.

Knuteson, J., C.E. Clapp, R.H. Dowdy, and W.E. Larson. 1988. Sewage sludge management. Land application of municipal sewage sludge to a terraced watershed in Minnesota. Minn. Agric. Exp. Sta. Misc. Publ. No. 56-1988. St. Paul.

Larson, W.E., C.E. Clapp, R.H. Dowdy, and D.R. Linden. 1994. Rosemount watershed study on land application of municipal sewage sludge. p. 123–126. *In* C.E. Clapp et al. (ed.) Sewage sludge: Land utilization and the environment. SSSA Misc. Publ. SSSA, Madison, WI (this publication).

Linden, D.R., C.E. Clapp, and R.H. Dowdy. 1983. Hydrologic and nutrient management aspects of municipal wastewater and sludge utilization on land. p. 79–101. *In* A.L. Page et al. (ed.) Utilization of municipal wastewater and sludge on land. Univ. of California, Riverside.

Stark, S.A., and C.E. Clapp. 1980. Residual nitrogen availability from soils treated with sewage sludge in a field experiment. J. Environ. Qual. 9:505–512.

21 Twenty Years of Trace Metal Partitioning on the Rosemount Sewage Sludge Watershed

R. H. Dowdy
C. E. Clapp
D. R. Linden

USDA-Agricultural Research Service
University of Minnesota
St. Paul, Minnesota

W. E. Larson
T. R. Halbach

Department of Soil Science
University of Minnesota
St. Paul, Minnesota

R. C. Polta

Metropolitan Waste Control Commission
St. Paul, Minnesota

Land application of sewage sludge has been studied intensively in the USA for the past 20 yr. Interest evolved through concerns for human health, environmental quality, rising costs of disposal, the philosophy of recycling, and recognition of sludge as a plant nutrient source. In spite of extensive research activity associated with land application of sludge, very few long-term, field-scale studies exist where sludge has been utilized as the sole nitrogen (N) source for crop production.

For 20 yr municipal sewage sludge has been applied annually at N utilization rates on a 16-ha watershed to determine if a continuous corn (*Zea mays* L.) farming system is sustainable and without environmental degradation. Previously, Dowdy and colleagues reported that: (i) sludge-borne trace metals had not enriched soil water extracted at 0.6 m and that cadmium (Cd), nickel (Ni), and lead (Pb) concentrations in runoff waters were not affected by sludge applications (Dowdy et al., 1987); (ii) sludge-borne Cd, chromium (Cr), copper (Cu), and zinc (Zn) migrated only into the 0.15- to 0.30-m soil layer directly below the zone of incorporation (Dowdy et al., 1989); and (iii) corn stover and grain did not accumulate sludge–borne Cd, Cr, Cu, Ni, or Pb during 17 yr of continuous corn production with annual sewage sludge applications (Dowdy et

Sewage Sludge: Land Utilization and the Environment. SSSA Miscellaneous Publication.

al., 1991). The useful life of any application site will be determined by, among other things, the quantities of sludge–borne metals added and their partitioning within the soil–plant continuum, the basis of this study.

MATERIALS AND METHODS

The 16-ha watershed consists of parallel terraces spaced 40 m apart with slopes ranging from 2 to 10% (Larson et al., 1978). The soil consists of 0.6 to 2.4 m of silt loam loess overlaying compacted glacial till. Surface soil pH ranged from 6.5 to 7.2. The area was cropped to continuous corn. Crop management has been reported by Duncomb et al. (1982) and Knuteson et al. (1988).

Digested sludge was applied annually by injection in the fall after corn harvest. A separate area (control) did not receive sludge, but was fertilized with comparable additions of mineral fertilizer. Cumulative sludge loading for the 20-yr period was 224 Mg ha^{-1}. Sludge Cd, Cu, Ni, Pb, and Zn concentrations were below the pollutant concentrations set by the U.S. Environmental Protection Agency (1993) for limiting total metal loading on a given site. Total sludge and metal loadings over time are given in Table 21-1. Chromium concentrations in the sludge were high, variable, and decreased dramatically after Year 10; but, since Cr is insoluble in soil systems, it served as an excellent tracer for following particulate movement in and through the watershed. The 2 *M* HCl sewage sludge ash (450°C) extracts were analyzed for metals by atomic absorption spectroscopy using deuterium-lamp background correction.

Metal composition of plant tissues were determined in a manner similar to that for sewage sludge. Soil and sediment samples were equilibrated overnight with 1 *M* HCl, and the supernant solution was analyzed spectroscopically. Water samples were acidified (pH < 1), concentrated, dried, ashed at 450°C, extracted with 2 *M* HCl, and analyzed for trace metals spectroscopically.

Table 21-1. Cumulative sludge and trace metal loadings on corn terraces over time.

Year	Sludge applied	Trace metal					
		Zn	Cu	Cd	Ni	Pb	Cr
	Mg ha^{-1}	kg ha^{-1}					
5	52[†]	45	35	0.3	0.9	15	195
10	130	115	95	0.8	2.8	35	875
17	164	145	120	1.0	4.0	42	975
19	205	175	135	1.2	4.9	49	1045

[†]Expressed on a dry weight basis, 105°C.

It was not possible to analyze the data statistically because of lack of true replication of the control area. Because trace metal concentrations in water, plant tissues, and soil were low, differences of less than two times are not considered meaningful.

RESULTS AND DISCUSSION

Cadmium, Ni, and Pb concentrations in runoff were not affected by 10 yr of sludge applications (Table 21-2). Cadmium concentrations never exceeded 2 $\mu g\ L^{-1}$. Sludge applications did enhance Cu concentrations in snowmelt runoff. These Cu levels were always 10 times less than U.S. Public Health Service drinking water standards. Slight, but elevated, Zn levels were observed in runoff waters. In spite of massive Cr loading (875 kg ha^{-1} in first 10 yr), Cr losses were low and only associated with sediment losses from corn areas, where sediments from early spring storms contained 10 times as much Cr as sediments from control areas. Sediment runoff, however, was minimal due to the terrace system; hence, only a very small fraction of applied Cr left the landscape. Evidence for trace metal migration to a 0.6-m soil depth did not exist during the first 5 yr. Consistently, higher concentrations of sodium (Na) and calcium (Ca) were observed in soil water extracted from sludge-treated areas at 0.6- and 1.5-m depths, demonstrating that water and metal cations were percolating through the soil profile.

Stover and grain did not have increased concentrations of sludge-borne Cd, Cr, Cu, Ni, or Pb (Table 21-3) during 19 yr of continuous corn production with annual sewage sludge applications as a plant nutrient source. Sludge-borne Zn did increase plant-Zn, reaching a mean concentration of 60 mg kg^{-1} in stover tissue for the 6- to 10-yr time interval. After 10 yr, tissue-Zn concentrations appeared to plateau (Fig. 21-1) as hypothesized by Corey et al. (1987) and observed by Bidwell and Dowdy (1987). The overall lack of sludge-borne metal accumulation by corn in our farming system supports the *clean sludge* concept (Chaney, 1989). This concept embodies the hypothesis that low metal sludges can be utilized as a nutrient source in farming systems without concern for significantly elevated trace metal levels in the growing crop. To a much lesser extent, the same trend was observed for Zn levels in corn grain and agrees with the broader observation that greatest metal accumulations occur in vegetative tissues.

Sludge-borne Cd, Cr, Cu, and Zn moved into the 0.16- to 0.3-m soil layer directly below the zone of incorporation by year 10 (Table 21-4). Because these increases were at least three times greater than values from the control area, the higher concentrations cannot be explained by variations in tillage depth (nominally 0.15 m). Nickel and Pb remained within the zone of incorporation. After 17 yr of sludge use, some movement of Cr and Cu into the 0.45- to 0.6-m layer occurred. In the case of Cr, this is the result of massive loading (975 kg Cr ha^{-1}). Finally, trace metals did not leach into soil beneath the lagoon following 15 yr of sludge storage. Soil pH below the base of the lagoon was >8.

Table 21-2. Metal concentrations in runoff water from corn terraced areas after 5 and 10 yr of annual sludge applications.

Year	Date	Treatment	Runoff	Zn	Cu	Cd	Ni	Pb	Cr
			m	μg L^{-1}					
5	Snowmelt	Control	0.021	22	10	<1	<10	<15	11
		Sludge	0.036	46	72	<1	<10	<15	33
	Rainfall†	Control	0.021	19	10	<2	4	<12	3
		Sludge	0.039	27	18	<2	9	<12	5
10	Snowmelt	Control	---	16	4	<1	4	<15	26
		Sludge	---	38	10	<1	<3	<15	21
	Rainfall‡	Control	0.028	10	9	<1	5	<15	15
		Sludge	0.031	20	11	<1	5	<15	16

†Runoff event on 7 April 1978.
‡Runoff event on 9 May 1983.

Table 21-3. Mean concentrations of trace metals in corn tissues over time.

Year	Treatment	Zn	Cu	Cd	Pb	Ni	Cr
		------------------------- mg kg^{-1} -------------------------					
				Stover			
1–5	Control	16.1†	5.0	0.14	1.9	0.7	1.0
6–10		17.7	6.0	0.10	0.9	0.8	1.4
11–19		18.0	8.4	0.16	0.9	0.7	0.9
1–5	Sludge	23.0	5.8	0.16	1.6	0.8	1.0
6–10		60.2	7.2	0.18	1.0	1.2	2.8
11–19		46.5	7.0	0.18	0.8	0.6	1.4
				Grain			
1–5	Control	21.9	1.7	0.07	0.6	0.4	0.2
6–10		23.2	2.4	0.03	0.3	0.3	0.2
11–19		20.0	3.2	0.29‡	0.4	0.4	0.2
1–5	Sludge	26.9	1.6	0.07	0.7	0.3	0.3
6–10		31.9	1.8	0.03	0.3	0.3	0.2
11–19		26.0	3.2	0.31‡	0.5	0.3	0.2

†Expressed on a 65°C weight basis.
‡May be a result of use of different hybrid varieties.

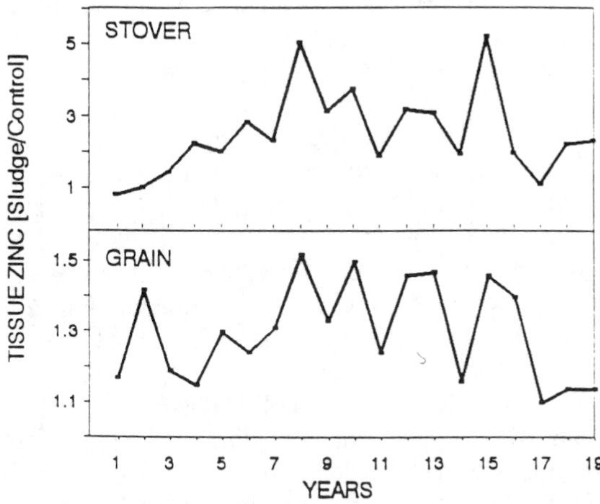

Fig. 21-1. Relative Zn concentration in corn stover and grain as a function of time with annual sludge applications.

Table 21-4. Extractable (1.0 *M* HNO_3) metals as a function of soil depth from corn terraces after 10 and 17 yr of sludge applications.

Depth	Zn	Cu	Cd	Ni	Pb	Cr
m			mg kg^{-1}			
			Control			
0.00–0.15	8/7†	3/3	0.10/0.20	4/5	8/6	5/4
0.16–0.30	8/7	5/3	0.10/0.17	5/7	12/5	5/3
0.31–0.45	-/7	-/3	-/0.17	-/7	-/4	-/2
0.46–0.60	-/7	-/3	-/0.17	-/6	-/3	-/2
			Sludge			
0.00–0.15	46/50	52/52	0.33/0.55	6/7	24/18	200/200
0.16–0.30	27/39	26/41	0.26/0.52	6/7	15/15	100/157
0.31–0.45	-/20	-/19	-/0.38	-/7	-/ 9	-/ 61
0.46–0.60	-/12	-/10	-/0.36	-/7	-/ 5	-/ 24

†10 yr/17 yr metal concentrations expressed on a 105°C weight basis.

REFERENCES

Bidwell, A.M., and R.H. Dowdy. 1987. Cadmium and zinc availability to corn following termination of sewage sludge applications. J. Environ. Qual. 16:438–442.

Chaney, R.L. 1989. Scientific analysis of proposed sludge rule. Biocycle 30:80–85.

Corey, R.B., L.D. King, C.Lue-Hing, D.S. Fanning, J.J. Street, and J.M. Walker. 1987. Effects of sludge properties on accumulation of trace elements by crops. p. 25–51. *In* A.L. Page et al. (ed.) Land application of sewage sludge. Lewis Publ. Chelsea, MI.

Dowdy, R.H., A.M. Bidwell, and C.E. Clapp. 1987. Trace metal movement in an agricultural watershed. p. 384–386. *In* Proc. 6th Int. Conf. on Heavy Metals in the Environ. New Orleans, LA 15–18 Sept. 1987. CEP Consultants, Edinburgh, Scotland.

Dowdy, R.H., C.E. Clapp, D.R. Linden, T.R. Halbach, and R.C. Polta. 1991. The progressive influence of 17 years of annual sewage sludge applications on trace metal accumulation. p. 379–382. *In* Proc. 8th Int. Conf. on Heavy Metals in the Environ. Edinburgh, Scotland. 16–20 Sept. 1991. CEP Consultants, Edinburgh, Scotland.

Dowdy, R.H., M.S. Dolan, C.E. Clapp, and W.E. Larson. 1989. Trace metal distributions in a loess soil following 10 years of sludge applications. p. 95–98. *In* Proc. 7th Int. Conf. on Heavy Metals in the Environ. Geneva, Switzerland. 11–15 Sept. 1989. CEP Consultants. Edinburgh, Scotland.

Duncomb, D.R., W.E. Larson, C.E. Clapp, R.H. Dowdy, D.R. Linden, and W.K. Johnson. 1982. The effects of liquid sewage sludge application on crop yield and water quality in a closed watershed. J. Water Pollut. Control Fed. 54:1185–1193.

Knuteson, J., C.E. Clapp, R.H. Dowdy, and W.E. Larson. 1988. Sewage sludge management. Land application of municipal sewage sludge to a terraced watershed in Minnesota. Minnesota Agric. Exp. Stn. Misc. Publ. 56–1988. St. Paul.

Larson, R.E., J.A. Jeffery, W.E. Larson, and D.R. Duncomb. 1978. A closed watershed for applying municipal sludge on crops. Trans. ASAE 21:124–128.

U.S. Environmental Protection Agency. 1993. Standards for the use or disposal of sewage sludge; Final rules. 40 CFR 257, 403, and 503. 19 Feb. 1993. Federal Register 58(32):9248–9415. U.S. Gov. Print. Office, Washington, DC.

SECTION V

REGULATIONS

22 Where Do We Stand on Regulations?

R. K. Bastian

Office of Wastewater Enforcement and Compliance
U. S. Environmental Protection Agency
Washington, DC

On 19 February 1993, the final 40 CFR Part 503 *Standards for the Use or Disposal of Sewage Sludge* and revisions to the 40 CFR Parts 122, 123, and 501 *NPDES Permit Regulations; State Sludge Management Program Requirements* were published in the Federal Register (58 FR [32]:9248-9415). The new U. S. Environmental Protection Agency (USEPA) rules address beneficial use practices involving land application as well as surface disposal and incineration of sewage sludge (or BIOSOLIDS). They affect generators, processors, users, and disposers of sewage sludge, both public and privately owned treatment works treating domestic sewage (including domestic septage haulers and non-dischargers), facilities processing or disposing of sewage sludge, and the users of sewage sludge and products derived from sewage sludge.

The Part 503 regulation addresses the use and disposal of only the sludge, including domestic septage, derived from the treatment of domestic wastewater. It does not apply to materials such as grease trap residues or other nondomestic wastewater residues pumped from commercial facilities, sludges produced by industrial wastewater treatment facilities, or grit and screenings from publicly owned treatment works (POTWs). Facilities which dispose of sewage sludge in municipal solid waste (MSW) landfills or use processed sewage sludge as a cover material are required to utilize facilities which comply with the requirements of the 40 CFR Part 258 MSW landfill rules issued in October 1991 (56 FR [196]:50978-51119].

Sewage sludge permitting requirements apply to *Treatment Works Treating Domestic Sewage* (TWTDS), i.e., facilities that generate, treat, or provide disposal of sewage sludge, including non-discharging and sludge-only facilities. A TWTDS must apply for a federal sewage sludge permit from the USEPA (or an approved state sludge program) if it manages a sewage sludge that is ultimately subject to Part 503, that is, it is land applied, placed on a surface disposal site, incinerated, or sent to a municipal solid waste landfill. TWTDS with existing NPDES permits get to apply for a sewage sludge permit as part of their next NPDES permit renewal application, while facilities without

Sewage Sludge: Land Utilization and the Environment. SSSA Miscellaneous Publication.

existing NPDES permits were required to submit limited screening information (but not necessarily a full permit application) by 19 Feb. 1994. While disposal facilities such as sewage sludge incinerators and surface disposal sites are clearly TWTDS and are required to apply for permits, the TWTDS definition does not extend automatically to areas such as farm land where sewage sludge is beneficially used; only under an unusual situation (such as a clear potential risk to human health and the environment) would the permitting authority designate such an area as a TWTDS. The permitting authority has the flexibility to cover both the generator and the treatment, use, or disposal facility in one permit or in separate permits (including general permits).

The USEPA will work closely with the states to encourage their adoption of sewage sludge management programs that can be approved to carry out the federal program and avoid the need for separate USEPA permits, compliance monitoring and enforcement activities. Until a state applies for and is approved to carry out a delegated program, however, all TWTDS in the state will be dealing directly with their USEPA Regional Office regarding federal permits, compliance monitoring and enforcement issues associated with the implementation of the Part 503 requirements, in addition to dealing with their state regulatory authorities and requirements.

By statute, compliance with the Part 503 standards is required within 12 months of publication of the new regulation (i.e., 19 Feb. 1994). If pollution control facilities need to be constructed to achieve compliance, then compliance is required within 2 yr (i.e., 19 Feb. 1995) of publication. Under the regulation, compliance with the monitoring and recordkeeping requirements was required within 150 d (i.e., 20 July 1993) of publication.

For the most part the Part 503 regulation is written to be self implementing, which means that citizen suits or USEPA can enforce the regulation even before permits are issued. As a result, treatment works must start monitoring and keeping records of sludge quality (and in many cases land appliers must start keeping records of loading rates and locations receiving sewage sludge), and must comply with pollutant limits and other technical standards, even in the absence of a federal permit.

Part 503 is organized into the following subparts: general provisions, land application, surface disposal, pathogens and vector attraction reduction, and incineration. Subparts under each of the use–disposal practices generally address: applicability, general requirements, pollutant limits, management practices, operational standards, frequency of monitoring, recordkeeping, and reporting requirements.

Under Part 503, Land Application includes all forms of applying sewage sludge to the land for beneficial uses at agronomic rates (rates designed to provide the amount of N needed by the crop or vegetation grown on the land while minimizing the amount that passes below the root zone). These uses include: application to agricultural land such as fields used for the production of food, feed and fiber crops, pasture and range land; non-agricultural land such as forests; disturbed lands such as mine spoils, construction sites and gravel pits; public contact sites such as parks and golf courses; and home lawns and gardens. The distribution and marketing of sewage sludge-derived materials such as composted, chemically-stabilized or heat-dried products is also addressed under land application, as is land application of domestic septage.

The rule applies to the entity who prepares sewage sludge for land application (including generating or deriving a material from sewage sludge), or who applies sewage sludge to the land. These parties must obtain and provide subsequent users or disposers the information necessary to comply with the rule. For example, the person who prepares bulk sewage sludge that is land applied must provide the person who applies it to the land all information necessary to comply with the rule, including the total N concentration of the sewage sludge.

The regulation establishes two levels of sewage sludge quality with respect to 10 heavy metal [arsenic (As), cadmium (Cd), chromium (Cr), copper (Cu), lead (Pb), mercury (Hg), molybdenum (Mo), nickel (Ni), selenium (Se) and zinc (Zn)] concentration limits -- pollutant Ceiling Concentrations and Pollutant Concentrations[1] (high quality sewage sludge), and two loading rate based limits, Cumulative Pollutant Loading Rates and Annual Pollutant Loading Rates (see Table 22-1); two levels of quality with respect to pathogen densities -- Class A and Class B (see Table 22-2); and two types of approaches for meeting vector attraction reduction, sewage sludge processing or the use of physical barriers at the land application or surface disposal site (see Table 22-3).

Under the Part 503 regulation, fewer restrictions are imposed on the use of higher quality sewage sludge. To qualify for land application, sewage sludge or material derived from sewage sludge must meet at least the pollutant Ceiling Concentration limits, Class B requirements for pathogens, and vector attraction reduction requirements. Cumulative Pollutant Loading Rates are imposed on bulk sewage sludges that meet the pollutant Ceiling Concentration limits but not the Pollutant Concentration limits; Annual Pollutant Loading Rates are to be imposed by specifying appropriate annual sludge loading rates on labels or handouts for these sludges when marketed in a bag or similar container. In all cases the minimum frequency of applicable monitoring, recordkeeping, and reporting requirements (see Table 22-4) must be met. A number of general requirements and management practices apply to sewage sludges that are land applied (see Table 22-5) with the exception of exceptional quality sewage sludge or derived material that meet the three quality requirements, the Pollutant Concentration limits, Class A pathogen requirements, and one of the vector attraction reduction sewage sludge processing requirements. In most cases, Exceptional Quality sewage sludge is no longer subject to the Part 503 requirements once it has been demonstrated and certified to meet the three requirements and passes on to the user for land application or producing other products. Based upon the results of the National Sewage Sludge Survey (NSSS) published in November 1991 (U.S. Environmental Protection Agency, 1990;

---

[1]A 25 Feb. 1994 *Federal Register* notice deleted the pollutant concentration limit (but not the ceiling concentration limit), the cumulative pollutant loading rate, and the annual pollutant loading rate for Mo until the USEPA reevaluates the issue based on information received from litigants. New values (which are expected to be higher, less stringent values) are to be proposed for public comment within 1 yr.

Table 22-1. Composition of sewage sludge vs Part 503 Ceiling and High Quality pollutant concentrations; cumulative and annual loading rates.

	Part 503 numeric criteria				NSSS results[†]	
	Table number					
	1	2	3	4		
Pollutant	Ceiling conc. limits[‡]	Cumulative pollutant loading	High quality pollutant conc. limits[§]	Annual pollutant loading rates	Median values[¶] (statistical basis)	
					Normal	Maximum
	mg kg^{-1}	kg ha^{-1}	mg kg^{-1}	kg ha^{-1} yr^{-1}	mg kg^{-1}	mg kg^{-1}
Arsenic	75	41	41	2.0	5	5
Cadmium	85	39	39	1.9	7	4
Chromium	3000	3000	1200	150	40	39
Copper	4300	1500	1500	75	463	456
Lead	840	300	300	15	106	76
Mercury	57	17	17	0.85	4	2
Molybdenum	75	18[#]	18[#]	0.90[#]	11	5
Nickel	420	420	420	21	29	18
Selenium	100	100	36	5.0	5	3
Zinc	7500	2800	2800	140	725	755

[†]USEPA, Nov. 1991. Summary of the National Sewage Sludge Survey (NSSS).
[‡]Absolute values.
[§]Monthly averages.
[¶]Pollutant concentrations for samples below the detection limit were incorporated into the estimates through the maximum likelihood procedure for multiple censor points to produce a better estimate than procedures that substitute either zero or the detection limit for nondetect samples.
[#]Mo values deleted on 25 Feb. 1994. New values to be proposed within 1 yr.

Table 22-2. Part 503 pathogen (indicator organism) density limits for Class A and Class B sewage sludges.

Classification	Fecal coliforms		Salmonella spp.
Class A[†]	<1000 MPN[‡] g^{-1} TS[‡]	-or-	< 3 MPN 4g^{-1} TS
Class B	<2 000 000 MPN g^{-1} TS		
	- or -		
	<2 000 000 CFU[‡] g^{-1} TS		

[†]In addition, density limits of >1 PFU 4g^{-1} TS for enteric virus and >1 4g^{-1} TS for viable helmith ova are included for evaluating sludge treatment processes that cannot meet specific operational requirements (i.e., time/temperature/pH relationships) specified in the rule.
[‡]MPN, most probably number; TS, total solids; CFU, colony forming unit; PFU, plaque forming unit.

Table 22-3. Vector attraction reduction alternatives.

Processing Alternatives

1. Aerobic or Anaerobic Digestion that achieves a ≥38% reduction in volatile solids (VS) measured as the difference in the raw sewage sludge prior to stabilization and the treated sewage sludge ready for use or disposal.
2. Anaerobic Digestion (if 38% VS reduction cannot be met) – demonstrated by further digesting a portion of the digested sewage sludge in a bench scale unit for an additional 40 days at 30 to 37°C or higher and achieving a further VS reduction of <17%.
3. Aerobic Digestion (if 38% VS reduction cannot be met) – demonstrated by further digesting a portion of the digested sewage sludge with a solids content of ≤2% in a bench scale unit for an additional 30 days at 20°C and achieving a further VS reduction of <15%.
4. Aerobic Digestion: Specific Oxygen Uptake Rate (SOUR) is ≤1.5 mg O_2 h^{-1} g^{-1} of total solids (TS) at 20°C.
5. Aerobic Processes: (e.g., composting) temperature is kept at >40°C for at least 14 days and the average temperature during this period is greater than 45°C.
6. Alkaline Stabilization: pH is raised to at least 12 by alkali addition and, without the addition of more alkali, remains at 12 or higher for 2 h and then at 11.5 or higher for an additional 22 h.

7.&8. Drying: TS is ≥75% when the sewage sludge does not contain unstabilized primary solids and ≥90% when unstabilized primary solids are included. Blending with other materials is not allowed to achieve the total solids percent.

Physical Barrier Alternatives

9. Injection: Liquid sewage sludge (or domestic septage) is injected beneath the surface with no significant amount of sewage sludge present on the surface after 1 h; sewage sludges that are Class A for pathogen reduction shall be injected within 8 h of discharge from the pathogen reduction process.
10. Incorporation: Sewage sludge (or domestic septage) that is land applied or placed in a surface disposal site shall be incorporated into the soil within 6 h of application; sewage sludge that is Class A for pathogen reduction shall be incorporated within 8 h of discharge from the pathogen reduction process.

Alternative for Surface Disposal or Septage Only

11. Surface Disposal Daily Cover: Sewage sludge or domestic septage placed in a surface disposal site shall be covered with soil or other material at the end of each operating day.

Alternative for Septage Only

12. Domestic Septage Treatment: The pH of domestic septage is raised to 12 or higher by alkali addition, and without the addition of more alkali, remains at 12 or higher for 30 minutes. This alternative is applicable to domestic septage applied to agricultural land, forest or reclamation sites or placed in a surface disposal site.

Table 22-4. Minimum frequency of monitoring, recordkeeping and reporting requirements.

Monitoring Frequency

Sewage sludge amounts	Monitoring frequency†
dry Mg yr^{-1}	
>0 to <290	once per year
290 to <1500	once per quarter
1500 to <15 000	once per 60 days
≥15 000	once per month

Record Keeping‡

Generators–Preparers: shall develop information and retain records:
- on the concentration of each chemical pollutant regulated under Part 405;
- certification (based on results of required periodic sampling–analysis) that the material meets the applicable pollutant concentration criteria; and
- certify that applicable pathogen reduction and vector attraction reduction requirements have been met.

Appliers: shall develop information and retain records:
- description of how the applicable management practices and site restrictions have been met for each application site;
- for sewage sludges limited by cumulative loading limits, keep records indefinitely of the cumulative amount of each pollutant applied to each site, information of the location and size of each site, date and time of applications, etc.; and
- certification that vector attraction reduction requirements have been performed in accordance with 503 if using injection or soil incorporation.

Reporting Frequency

Annual reporting is required of all Class I sewage sludge management facilities (i.e., the ≈1600 pretreatment POTWs§ and ≈400 other designated TWTDS§ such as sewage sludge only facilities) and other major POTWs – those with a design flow ≥1 MGD§ or serving a population of ≥10 000 people. In addition, for sites where recordkeeping is required, the same group of facilities shall report annually when any cumulative metal loading reaches 90% of the allowed Cumulative Pollutant Loading Rates (503 Table 2 values).

†The permitting authority may impose more frequent monitoring requirements on permittees; in addition, after two years of monitoring at these frequencies, the permitting authority may allow the monitoring frequencies to be reduced to no less than once per year.

‡Recordkeeping requirements vary with the end use of the sewage sludge or derived material. Except as noted records must be kept for 5 yr.

§POTW, publicly owned treatment works; TWTDS, treatment works treating domestic sewage; MGD, million gallons per day.

Table 22-1), USEPA anticipates that a large percentage of the sewage sludges currently being produced will be capable of meeting the high quality pollutant concentration limits.

Under Part 503, Surface Disposal addresses the disposal of sewage sludge and septage on land, including sludge-only monofills; dedicated surface disposal application sites (where sludge pollutants are applied at higher than the Cumulative Pollutant Loading Rates or sludge is applied at higher than agronomic rates allowed under land application); piles or mounds and impoundments or lagoons where sludge is placed for final disposal. It is not intended to include the placement of sewage sludge in similar locations for storage or treatment; however, the facility operator will need to provide an adequate justification concerning why it is being stored for longer than 2 yr.

For surface disposal, the regulation establishes requirements for active sewage sludge disposal units with and without liners and leachate collection systems. Concentration limits for three heavy metals (As, Cr, and Ni) apply to sewage sludges placed in active sewage sludge disposal units without a liner and leachate collection system, but no concentration limits apply to sewage sludge placed in a unit with a liner and leachate collection system. Specific management practice requirements address such areas as the location of surface disposal sites, control of surface runoff, monitoring ground water and for methane gas production, and restrictions on crop production, grazing and public access. A provision allowing for the establishment of site-specific limits and management practices for surface disposal sites is provided.

Part 503 establishes minimum monitoring frequencies that are based upon the annual amount of sewage sludge used or disposed; maintenance of records (in most cases for a minimum of 5 yr) regarding such information as sludge quality, application sites, application dates, various certification statements and descriptions of management practices, and pathogen and vector attraction reduction measures used. Annual reporting is required only for Class I sludge management facilities (the ≈1600 pretreatment POTWs and an estimated 400 other facilities likely to be designated Class I TWTDS, and other POTWs with a design flow of ≥1 MGD or serving a population ≥10 000.

For the most part, Part 503 is a risk-based regulation designed to protect public health and the environment from reasonable worst case situations. Models were established to facilitate the evaluation of fourteen major pathways of pollutant exposure (e.g., sludge→soil→plant→human; sludge→soil→soil biota) associated with land application practices. Fifty pollutants of potential concern were evaluated using the available scientific data. Acceptable sludge pollutant concentration limits and loading rates were calculated based upon conservative assumptions and endpoints previously established as adequate to protect public health and the environment.

Concerns have been raised over some of the scientific data, assumptions and models used in developing the numerical limits. As required by the Clean Water Act (and several pending lawsuits), future rounds of rulemaking are expected to address these concerns and additional pollutants that may be present in sewage sludge. In materials submitted to the court during May 1993 in response to pending lawsuits on the final Part 503 regulation, USEPA provided

Table 22-5. General requirements and management practices for land application.[†‡]

General Requirements

- Bulk sewage sludge subject to cumulative pollutant loading rates shall not be applied to agricultural land, a forest, public contact or reclamation site if any of the cumulative pollutant loading rates have been reached.
- Preparers of bulk sewage sludge to be applied to agricultural land, a forest, public contact or reclamation site shall provide appliers written notification of the total N concentration (dry wt) in the bulk sewage sludge and other information necessary to comply with the 503 requirements.
- Appliers shall obtain information to comply with requirements, contact the permitting authority and others to determine if (and how much) material subject to cumulative loadings has been applied before.
- Appliers of bulk sewage sludge shall provide land application site owners or lease holders with notice and information necessary to comply with the 503 requirements.
- Preparers of bulk sewage sludge to be applied in a state other than the state in which the material is prepared shall provide written notice, prior to the initial application of the bulk material to a land application site by the applier, to the permitting authority for the state in which the bulk material is proposed to be applied.
- Appliers of bulk sewage sludge subject to cumulative loading rates shall provide written notice, prior to the initial application of bulk sewage sludge to a land application site by the applier, to the permitting authority for the state in which the bulk sewage sludge will be applied and the permitting authority shall retain and provide access to the notice.

Management Practices

- Bulk sewage sludge shall not be applied to the land if it is likely to adversely affect a threatened or endangered species listed under the Endangered Species Act or its designated critical habitat.
- Bulk sewage sludge shall not be applied to agricultural land, a forest, public contact or reclamation site that is flooded, frozen or snow-covered ground so that the sewage sludge enters a wetland or other waters of the USA except as provided in a permit issued under section 402 or 404 of the Clean Water Act.
- Bulk sewage sludge shall not be applied to agricultural land, a forest, public contact or reclamation site at above agronomic rates, with the exception of reclamation projects when authorized by the permitting authority.
- Bulk sewage sludge shall not be applied to agricultural land, a forest or reclamation site that is <10 m from waters of the USA unless authorized by the permitting authority.
- For sewage sludge sold or given away, either an appropriate label shall be affixed to the material's bag or container, or an information sheet containing specific information shall be provided to the receiver of the material for land application.

[†]The permitting authority may apply any or all of the general requirements or management practices to land application of bulk exceptional quality (EQ) sewage sludge or a product derived from an EQ sewage sludge on a case-by-case basis if determined needed to protect public health and the environment.

[‡]In addition, when sewage sludge that meets Class B pathogen reduction requirements, but not Class A, is applied to the land, site restrictions apply regarding waiting periods before harvesting crops, grazing animals, and allowing public access.

Table 22-6. Pollutants identified for Round II.[†]

- Aluminum
- Antimony
- Asbestos
- Barium
- Beryllium
- Boron
- Butanone (2-)
- Carbon disulfide
- Cresol
- Cyanides (soluble salts and complexes)
- Dichlorophenoxyacetic acid (2, 4-D)
- Dioxins/Dibenzofurans (all monochloro to octochloro congeners)
- Endosulfan-II
- Fluoride
- Manganese
- Methylene chloride
- Nitrate
- Nitrite
- Pentachloronitrobenzene
- Phenol
- Phthalate (bis-2-ethylhexyl)
- Polychlorinated biphenyls (co-planar)
- Propanone (2-)
- Silver
- Thallium
- Tin
- Titanium
- Toluene
- Trichlorophenoxyacetic acid (2, 4, 5-T)
- Trichlorophenoxypropionic acid
- Vanadium

[†]List of 31 pollutants identified in a Notice to the Court on 14 May 1993, that the USEPA intends to propose for regulation no later than 15 Dec. 1999, although the USEPA retains the discretion to either add or delete pollutants from this list.

a list of 31 additional pollutants that the Agency intends to further evaluate via research and risk assessment, and propose for regulation no later than 15 Dec. 1999 (see Table 22-6). Areas such as the long-term fate of some land applied pollutants in sewage sludge relative to plant uptake rates, surface runoff and ground water movement, and the potential impacts (both positive and negative) on wildlife and unmanaged ecosystems are ripe for further research due to the limited amount of field data currently available. Future attempts to make the pathogen control portion of the rule more risk-based will also require additional research efforts.

It has been the policy of the USEPA and the Federal government as a whole to encourage beneficial use practices when done in a manner that is protective of public health and the environment (U. S. Environmental Protection Agency/United States Department of Agriculture/Food and Drug Administration, 1981; U.S. Environmental Protection Agency, 1984; U.S. Environmental Protection Agency, 1991). However, to make the past efforts invested in developing appropriate land application practices (including those by the University of Minnesota and USDA-Agricultural Research Service at the Rosemount Agricultural Experiment Station) and the Part 503 requirements effective in encouraging more beneficial use of sewage sludge, additional efforts will be needed in many parts of the country to improve public acceptance of these practices. Activities being undertaken by the Water Environment Federation to promote BIOSOLIDS Recycling (Water Environment Federation, 1993) are a step in this direction. Continued efforts on the part of agricultural extension specialists, soil scientists, and agronomists will be needed to help assure farmers and their neighbors, food processors, state regulators, local politicians, environmentalists, and other special interest groups that treated sewage sludge (BIOSOLIDS) meeting USEPA's Part 503 standards can be safely and effectively applied to the land in a beneficial manner.

REFERENCES

U. S. Environmental Protection Agency/U. S. Department of Agriculture/Food and Drug Administration. 1981. Land application of municipal sewage sludge for the production of fruits and vegetables; A statement of federal policy and guidance. SW-905. USEPA, Office of Solid Waste. Washington, DC.

U. S. Environmental Protection Agency. 1984. EPA policy on municipal sludge management. Federal Register 49(114):24358–24359 (12 June 1984). U.S. Gov. Print. Office, Washington, DC.

U. S. Environmental Protection Agency. 1990. National sewage sludge survey: Availability of information and data, and anticipated impacts on proposed regulations. Federal Register 55(218):47210–47283 (9 Nov. 1990). U.S. Gov. Print. Office, Washington, DC.

U. S. Environmental Protection Agency. 1991. Interagency policy on beneficial use of municipal sewage sludge on federal land. Federal Register 56(138):33186–33188 (18 July 1991). U.S. Gov. Print. Office, Washington, DC.

Water Environment Federation. 1993. Biosolids - Too valuable to waste. Water environment technology. Water Environment Federation, Alexandria, VA.

SECTION VI

POSTER CONTRIBUTIONS

23 Effect of Lime-Cake Municipal Sewage Sludge on Corn Yield, Nutrient Uptake, and Soil Analyses

Christara M. Hormann

Department of Soil Science
University of Minnesota
St. Paul, Minnesota

C. E. Clapp
R. H. Dowdy

USDA-Agricultural Research Service
University of Minnesota
St. Paul, Minnesota

W. E. Larson
D. R. Duncomb
T. R. Halbach

Department of Soil Science
University of Minnesota
St. Paul, Minnesota

R. C. Polta

Metropolitan Waste Control Commission
St. Paul, Minnesota

Land application of sewage sludge has become an accepted alternative to alleviate problems associated with incineration and the impact on air and water quality. Municipal sewage sludge treated with hydrated lime and dewatered, forms a biologically stabilized sludge. A cake-like solid is produced containing ≈70 to 80% water, 30% lime as $CaCO_3$ equivalent, and a pH of >11.0. The addition of hydrated lime makes it uniquely different from other sewage sludges applied to Minnesota soils (Dowdy & Larson, 1975ab; Dowdy et al., 1978; Stark & Clapp, 1980; Duncomb et al., 1982). Researchers have shown that amending agricultural soils with lime-treated wastes would increase soil pH and calcium (Ca) concentration in crops, while potentially decreasing potassium (K) uptake

Sewage Sludge: Land Utilization and the Environment. SSSA Miscellaneous Publication.

(Soon et al., 1978ab; Logan & Chaney, 1983). Applying sewage sludge to calcareous or limed soils has also been shown to reduce toxic plant concentrations of trace metals (Bingham et al., 1979; Chang et al., 1980; Soon et al, 1980; Morel & Guckert, 1981). Our objectives in this experiment were: (i) to determine effects of lime-cake sewage sludge on corn (*Zea mays* L.) yield, on corn nutrient uptake and on soil chemical properties; and (ii) to determine effects of cations and trace metals from lime-cake sludge on cation balance in the soil and on plant metal contents.

MATERIALS AND METHODS

Site Location and Design

The study was conducted on a 0.2 ha site at the Rosemount Agricultural Experiment Station of the University of Minnesota. The site situated on a slope of 2 to 5% had a top soil depth of 30 to 45 cm and had been mapped as Port Byron silt loam (fine-silty, mixed, mesic Typic Hapludoll) formed on deep loess (>240 cm) over glacial till. Twelve, 0.01-ha plots (four treatments x three replicates) were arranged in a randomized block design on the site allowing for eight corn rows, 16 m in length at 0.75-m spacing on each plot (Larson et al., 1983).

Treatments were: (i) control (fertilized for maximum crop production); (ii) 15 dry Mg ha^{-1} of lime-cake sewage sludge per year (approached limit for yearly cadmium (Cd) application per Minnesota Pollution Control Agency, 1982); (iii) 30 dry Mg ha^{-1} of lime-cake sludge per year (approached limit for yearly nitrogen (N) application); and (iv) 60 dry Mg ha^{-1} of lime-cake sludge per year. Sewage sludge was applied to plot areas in the spring before the first crop and then each fall following harvest for five consecutive years (total of six applications). Sludge was applied using a manure spreader and then incorporated with a tractor-mounted rototiller. Samples were taken during spreading for each plot and composited for laboratory analysis.

Control plots received annual applications of 260 kg N ha^{-1} as NH_4NO_3 and 28 kg phosphorus (P) ha^{-1} as triple superphosphate. Potassium was applied to the control plots for the fifth growing season and to all plots in Years 7 and 8 at 90 kg K ha^{-1} as KCl.

Corn was planted in early May at 65 000 kernels ha^{-1}. Corn rootworm insecticide was applied at planting followed by a preemerge broadcast application of a herbicide for weed control. Corn was harvested in late September for yield and laboratory analyses along two 6-m sampling lines from each plot. The remaining corn fodder was removed from the plot area.

Soil samples were taken before initial sludge application and after harvest each fall. Plant and soil sampling continued for 4 yr following the sixth (final) sludge application.

Laboratory Analyses

Total solids of sludge samples were determined at 105°C. Total carbon (C) was measured by an automated determinator (Nelson & Sommers, 1982). Total N was determined by a semimicro-Kjeldahl procedure with a $CuSO_4$–Se digestion catalyst (Bremner & Mulvaney, 1982) and NH_4–N determined by distillation and titration. Samples for total P, K, Ca, iron (Fe), aluminum (Al), magnesium (Mg), sodium (Na), and trace metals were dry-ashed at 450°C for 24 h, equilibrated with 2.0 *M* HCl and the supernatant was analyzed by inductively coupled plasma emission or atomic absorption spectroscopy using background correction for copper (Cu), nickel (Ni), Cd, lead (Pb), and zinc (Zn). Soil pH and electrical conductivity (EC) were determined from the extract of a 1:1 soil to water mixture. Plant samples were oven dried at 65°C, ground, and analyzed for N, P and metals by the same methods as used for sludge.

RESULTS AND DISCUSSION

Sludge Application

Composition of lime-cake sewage sludge spread during a 6-yr period is shown in Table 23-1. Sludge was delivered to the field site as a cake material consisting of ≈25% solids. Calcium content was >14%, accounting for and contributing to the high pH and very low NH_4–N concentrations. The major N fraction was in the organic form. Phosphorus was about half the total N concentration, while sludge K concentrations were low. Concentration of Cd was relatively high.

Total sludge applied for the 6-yr period is shown in Table 23-2. The three sludge treatments received annual applications of about 20, 30, and 60 dry Mg ha^{-1} yr^{-1} for low, medium, and high rates, respectively. Total yearly Cd additions were 2, 3, and 6 kg ha^{-1} for low, medium, and high plots. Minnesota Pollution Control Agency (1982) sludge application guidelines for the Port Byron soil restricted yearly Cd rates to 2.2 kg ha^{-1} with a total site accumulation not to exceed 22 kg ha^{-1}. Sludge applications on the medium and high treatments exceeded the annual Cd rate, while five sludge applications on the high treatment exceeded the Cd site life.

Corn Yields

No significant differences (Duncan's multiple range test) were detected in corn grain and fodder yields due to lime-cake sludge application (Fig. 23-1). Control yields were equal to or greater than sludge yields for most years with greater differences occurring in the final 2 yr after sludge application ceased. Yields for all treatments were average to high for the southeastern area of Minnesota. During the third cropping season a wind and hail storm severely damaged the corn crop, substantially reducing yields. Corn fodder yields from the sludge treatments were reduced during the fifth, sixth, and seventh cropping years, possibly caused by reduced K uptake.

Table 23-1. Composition of lime-cake municipal sewage sludge for six annual applications.

Constituents[†]	Mean	Constituents[†]	Mean
	%		mg kg^{-1}
Total solids	24.8	Zn	1300
Total C	25.7	Cr	1000
Total N	2.98	Mn	700
NH_4–N	0.25	Cu	670
P	1.20	Pb	300
K	0.07	Ni	160
Ca	14.1	Cd	86
		B	18
Fe	3.6	Electrical conductivity	dS m^{-1} 4.3
Al	1.0		
Mg	0.5		
Na	0.1	pH	11.6

[†]Total solids based on 105°C dry weight. Other constituents based on percentage of total solids.

Table 23-2. Lime-cake sewage sludge and fertilizer applications for 6 yr.

Treatment	Application rate		$CaCo_3$ equiv.	Total N	Total P	Total K	Total Cd
	Annual	Total					
	Mg ha^{-1} yr^{-1}	Mg ha^{-1}	Mg ha^{-1} yr^{-1}	kg ha^{-1} yr^{-1}			
Control	0	0	0	260	28	90	0
Low	19.5	117	6.8	560	230	13	2.0
Medium	31.8	188	11.5	930	370	20	3.2
High	58.6	352	21.2	1720	700	39	5.9

Plant Analyses and Nutrient Uptake

Values for the major plant nutrients N and P for the lime-cake plots were within adequate ranges for normal corn growth. Corn fodder removed 30, 19, and 10% of the applied N from the low, medium, and high plots, respectively (Fig. 23-2), with 68% removal of the applied inorganic N fertilizer from the control plots. Removal of P by corn fodder was relatively small in comparison to the amount applied, with 11, 7, and 4% uptake for the low, medium, and high plots, respectively.

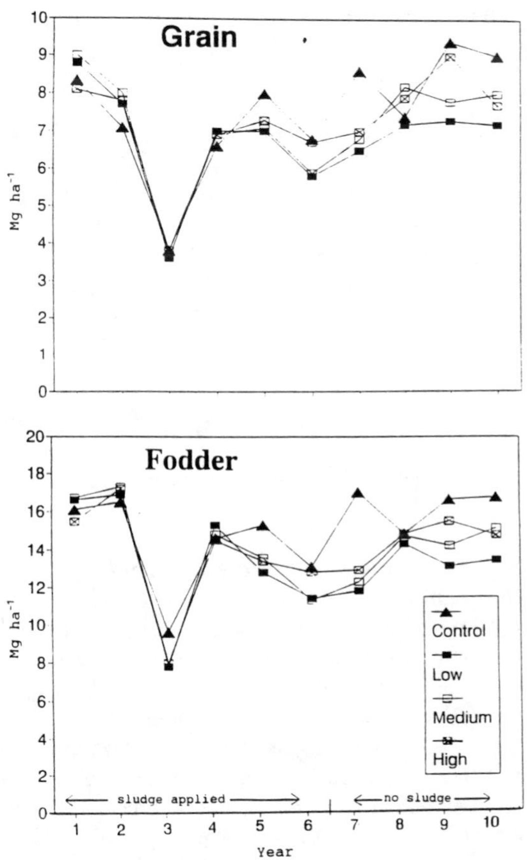

Fig. 23-1. Corn grain and fodder yields.

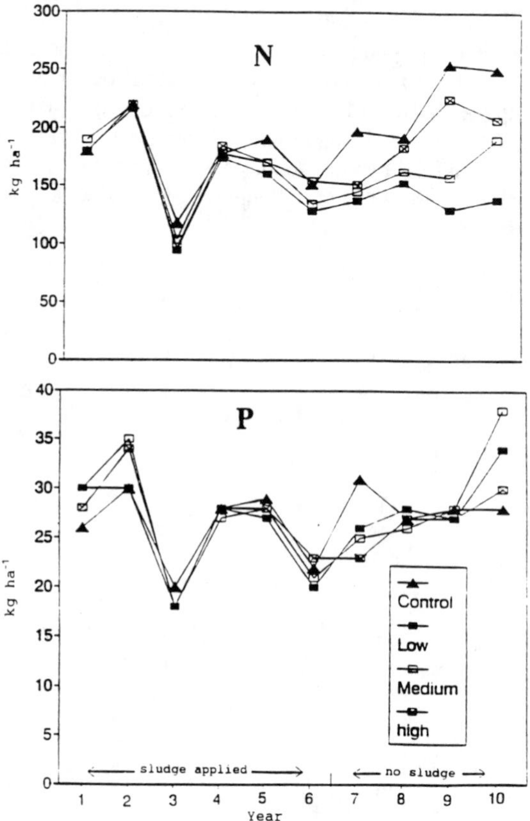

Fig. 23-2. Nutrient uptake of nitrogen and phosphorus.

Fodder K removed greatly exceeded the amount of K applied (data not shown). The high sludge treatment supplied approximately half the recommended K for corn production based on University of Minnesota Soil Testing Laboratory results. Competition with K by the Ca for colloidal exchange sites could pose availability problems for K uptake. Imbalances in Ca and K, soil pH, carbonates, neutral salts, organic matter, and CO_2 in the soil profile could have had an effect on corn uptake of K. All plots received K for the seventh and eighth growing seasons, slightly increasing corn yields.

Stover Cd increased significantly due to lime-cake sludge addition (Fig. 23-3), but not nearly as much as did the stover reported by Dowdy et al. (1981), receiving unlimed sludge from the same treatment plant. A decrease in Cd was observed in stover in years following final sludge application. Long-term observations show that plant availability of metals in sludge-treated soils either remain unchanged or is reduced with time after cessation of sludge application (Bidwell & Dowdy, 1987). Corn grain Cd was not increased by lime-cake sludge

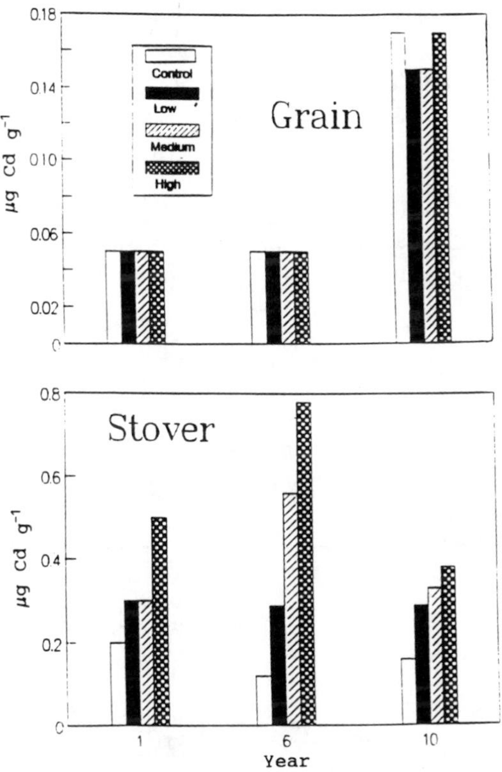

Fig. 23-3. Cadmium concentration in corn grain and stover.

additions (Fig. 23-3). Soon et al. (1980) also showed low corn grain Cd accumulation for application of lime-treated sludge. The slight increase of grain Cd for all treatments in Year 10 is probably due to a change in corn variety that took place during Years 7 to 10.

Corn tissue concentrations of other trace metals from the lime-cake sludge-treated plots were not significantly influenced by sludge addition (Duncan's multiple range test).

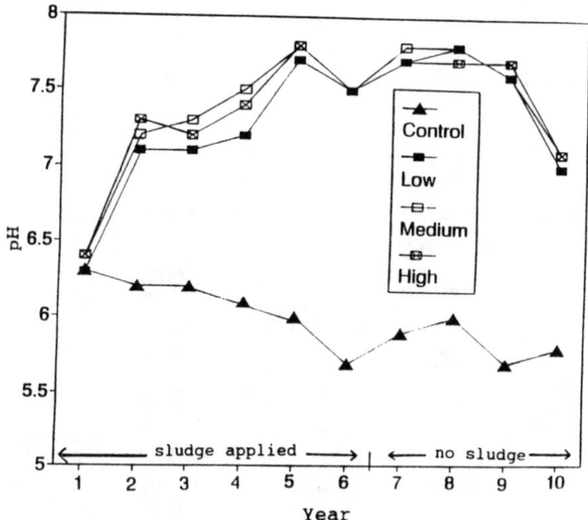

Fig. 23-4. Soil pH.

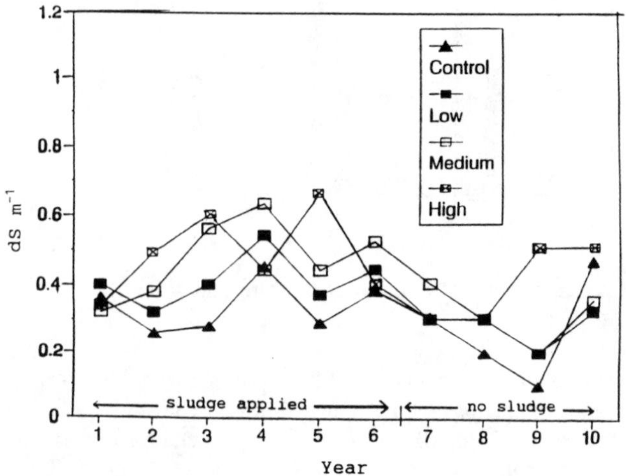

Fig. 23-5. Soil electrical conductivity.

Soil Analyses

Six sludge applications supplied a considerable amount of lime, greatly exceeding the liming requirements of the Port Byron soil. Nearly 17 Mg ha^{-1} of $CaCO_3$ at 15 cm or 15.4 $cmol_c$ kg^{-1} were needed to neutralize soil exchange acidity at the start of the experiment determined by the barium chloride-triethanolamine extraction method (Thomas, 1982). Soil pH in the surface 15 cm, after six cropping seasons with sludge treatment (Fig. 23-4), was significantly different from the control treatment, but did not differ between sludge treatments. After sludge applications ceased, pH decreased, but did not return to the value of the control plots.

Soil EC showed an increase due to lime-cake sludge application (Fig. 23-5). The greatest increase was for the medium and high treatments, primarily due to high Ca and Mg concentrations, since sludge Na and K values were very low. Electrical conductivity values of 0.3 to 0.6 dS m^{-1} for the high treatment were considerably below the 2.0 to 4.0 dS m^{-1} range for very slightly saline soils.

CONCLUSIONS

Corn grain and fodder yields obtained on the lime-cake sludge plots were not significantly different between treatments and were approximately equal to the control plots.

Nutrient uptake of major plant nutrients was in the range for normal corn growth, except for stover K. All sludge treatments were near K deficiency range, although the high application plots had greater stover-K concentrations. The high sludge treatment supplied 40 kg K ha^{-1} with the sludge.

Trace metal concentrations in lime-cake corn tissue remained within the range for normal growth. Slight increases in stover Zn and Cd were noted.

Soil pH increased significantly with lime-cake sludge addition, reducing the availability of trace metals and decreasing the potential for entering the food chain. Electrical conductivity increased with higher lime-sludge application rates due to soluble salts in the soil.

Results indicated that lime-cake municipal sewage sludge can be applied to agricultural land with no detrimental effect on the environment, while maintaining good crop yields.

REFERENCES

Bidwell, A.M., and R.H. Dowdy. 1987. Cadmium and zinc availability to corn following termination of sewage sludge applications. J. Environ. Qual. 16:438–442.

Bingham, F.T., A.L. Page, G.A. Mitchell, and J.E. Strong. 1979. Effect of liming an acid soil amended with sewage sludge enriched with Cd, Cu, and Zn on yield and Cd content of wheat grain. J. Environ. Qual. 8:202–207.

Bremner, J.M., and C.S. Mulvaney. 1982. Nitrogen-total. p. 595–624. *In* A.L. Page et al. (ed.) Methods of soil analysis. Part 2. 2nd ed. Agronomy

Monogr. 9. ASA and SSSA, Madison, WI.

Chang, A.C., A.L. Page, and F.T. Bingham. 1980. Re-utilization of municipal wastewater sludges--metals and nitrate. J. Water Pollut. Control. Fed. 53:237–245.

Dowdy, R.H., and W.E. Larson. 1975a. Metal uptake by barley seedlings grown on soils amended with sewage sludge. J. Environ. Qual. 4:229–233.

Dowdy, R.H., and W.E. Larson. 1975b. The availability of sludge-borne metals to various vegetable crops. J. Environ. Qual. 4:278–282.

Dowdy, R.H., W.E. Larson, J.M. Titrud, and J.J. Latterell. 1978. Growth and metal uptake of snap beans grown on sludge amended soils, a four year field study. J. Environ. Qual. 7:252–257.

Dowdy, R.H., P.K. Morphew, and C.E. Clapp. 1981. The relationship between the concentration of cadmium in corn leaves and corn stover grown on sludge amended soils. p. 466–477. *In* Proc. 4th Annual Madison Conf. on Municipal and Industrial Waste, Madison, WI. 28-30 Sept. 1981. Univ. of Wisconsin, Madison, WI.

Duncomb, D.R., W.E. Larson, C.E. Clapp, R.H. Dowdy, D.R. Linden, and W.K. Johnson. 1982. Effect of liquid wastewater sludge application on crop yield and water quality. J. Water Pollut. Control Fed. 54:1185–1193.

Larson, W.E., C.E. Clapp, R.H. Dowdy, D.R. Duncomb, and R.C. Polta. 1983. The effect of limed sludge on soil properties and maize production. p. F.1–F.12. *In* Utilization of sewage wastes on land. Research progress report. USDA-ARS, Univ. of Minnesota, St. Paul.

Logan, T.J., and R.L. Chaney. 1983. Utilization of municipal wastewater and sludge on land--metals. p. 235–326. *In* A.L. Page et al. (ed.) Proc. workshop utilization municipal wastewater and sludge on land. Denver, CO. 23-25 Feb. 1983. Univ. of California, Riverside.

Minnesota Pollution Control Agency. 1982. Pollution Control Agency Sewage Sludge Management Rules, Chapter 7040, Minnesota Pollution Control Agency, St. Paul.

Morel, J.L., and A. Guckert. 1981. Influence of limed sludge on soil organic matter and soil pH properties. p. 25–42. *In* G. Catroux et al. (ed.) The influence of sewage sludge application on physical and biological properties of soils. D. Reidel Publ. Co., Dordrecht, Holland.

Nelson, D.W., and L.E. Sommers. 1982. Total carbon, organic carbon, and organic matter. p. 539–594. *In* A.L. Page et al. (ed.) Methods of soil analysis. Part 2. 2nd ed. Agron. Monogr. 9. ASA and SSSA, Madison, WI.

Soon, Y.K., T.E. Bates, E.G. Beauchamp, and J.R. Moyer. 1978a. Land application of chemically treated sewage sludge: I. Effects on crop yield and nitrogen availability. J. Environ. Qual. 7:264–269.

Soon, Y.K., T.E. Bates, and J.R. Moyer. 1978b. Land application of chemically treated sewage sludge: II. Effects on plant and soil phosphorus, potassium, calcium, and magnesium and soil pH. J. Environ. Qual. 7:269–273.

Soon, Y.K., T.E. Bates, and J.R. Moyer. 1980. Land application of chemically treated sewage sludge: III. Effects on soil and plant heavy metal content. J. Environ. Qual. 9:497–504.

Stark, S.A., and C.E. Clapp. 1980. Residual nitrogen availability from soils treated with sewage sludge in a field experiment. J. Environ. Qual. 9:505–512.

Thomas, G.W. 1982. Exchangeable cations. p. 159–165. *In* A.L. Page et al. (ed.) Methods of soil analysis. Part 2. 2nd ed. Agron. Monogr. 9. ASA and SSSA, Madison, WI.

24 Sludge-Borne Salt Effects on Soybean Growth

G. H. Abd El-Hay

Faculty of Agriculture
Al-Azhar University
Nasr City, Cairo, Egypt

J. Scott Angle

Department of Agronomy
University of Maryland
College Park, Maryland

Sludge-borne salts may have a detrimental effect on the growth and yield of soybeans cultivated on sludge-amended land. A greenhouse pot experiment was conducted where soybeans [*Glycine max* (L.) merr., var. Jackson and the salt tolerant variety, Lee] were grown in soil amended with varying rates of anaerobically digested sludge or NaCl. Sewage sludge was added to soil at rates of 0, 200, 400, and 800 Mg ha^{-1} and NaCl was added to soil at rates of 0, 2.5, 5.0, 10, and 20 g kg^{-1}. Electrical conductivity of soil increased with increasing sludge addition to soil. At rates exceeding 200 kg ha^{-1}, plants exhibited symptoms of salt toxicity and growth and shoot nitrogen (N) were significantly reduced. At the control rate, growth of Jackson exceeded that of Lee, however, when sludge was added to soil, shoot weights were highest with the variety Lee. Shoot weight and N were higher at the 200 kg ha^{-1} sludge rate than that of the control. In a second experiment, sludge-amended soils were leached with distilled water to flush soluble salts from the mixture. Soybeans were then planted into the soil as before. When salts were removed from the mixture, growth and shoot N content were enhanced up to a sludge addition rate of 400 kg ha^{-1}. At 800 kg ha^{-1}, growth was equal to that of the unamended control. These results demonstrate that the soluble salt content of sludge-amended soil may restrict growth of soybeans when high rates of sludge are added to soil and that these adverse effects can be reduced if rainfall or irrigation water remove salts from the root zone prior to planting.

Sewage Sludge: Land Utilization and the Environment. SSSA Miscellaneous Publication.

25 Residual Concentration of Selected Heavy Metals in a Sewage Sludge–Amended Soil and Uptake by Coastal Bermudagrass

A. Garcia
A. S. Mangaroo

Department of Agriculture
Prairie View A&M University
Prairie View, Texas

Soil is often used to collect and purify water and dispose of anthropogenic wastes, including sewage sludge. Pollutants in the soil-water system result from all types of anthropogenic activity, agricultural, industrial, and residential land uses. In order to effectively manage urban solid wastes with the objective of maximum recovery of resources, rural land utilization for disposal is essential. In recent years, sewage sludge from treatment plants in neighboring suburban and densely populated Harris County of Texas is being utilized in rural Waller County on farmland. Not only is the sludge-treated land being used to fertilize forage, it is also being used in some instances for food crops. Economic and environmental considerations in the future are expected to result in substantial increases in the amounts of sewage sludge applied to cropland in the vicinity.

Sewage sludge is known to provide the soil with the heavy metals present; copper (Cu), iron (Fe), cobalt (Co), manganese (Mn), molybdenum (Mo), and zinc (Zn), that are valuable micronutrients. Other heavy metals present are lead (Pb), cadmium (Cd), chromium (Cr), nickel (Ni), mercury (Hg), and tin (Sn). These metals are known for their toxicity, and tend to accumulate in the soil and in plant tissues. The possibility exists of these elements entering the food chain. There also is the tendency of these chemical elements, organic constituents, and other soil constituents to move down the soil profile (Greenland & Hayes, 1981).

In recent years there have been numerous studies of the accumulation, or uptake of heavy metals by crop plants growing in sludge-treated soil (Chaney, 1973; Hinesly et al., 1977; Leeper, 1978; Day & Thompson, 1986; King & Giordano, 1986. While King and Giordano (1986) reported on the uptake of heavy metals by forages, there is still insufficient data on forage uptake of heavy metals from sludge-treated soils.

Sewage Sludge: Land Utilization and the Environment. SSSA Miscellaneous Publication.

Stevenson (1987) indicated the need for research to establish the extent of health risks associated with the use of agricultural chemicals and organic wastes on farm land and the pollution of soils and natural waters with anthropogenic organic chemicals. The five-year research plan of the Texas Agricultural Experiment Station under environmental quality has identified: *Land Treatment and Disposal of Agricultural, Industrial, and Municipal Wastes*, and the *Assessment of Soil Pollutants in Pathways of Agricultural Food Chains*, as priority research areas for the state of Texas. Research is needed to quantify the relationship between crop growth, crop quality, and critical soil–water properties of the sludge-treated soils of this area. The objectives of this study were to determine: (i) concentration of the heavy metals Zn, Cd, Cu, Pb, and Ni in coastal bermudagrass [*Cynodon dactylon* (L.) Pers.] forage produced with and without HouActinite (CDR Environmental, Inc., Houston, TX) applied to soil; (ii) concentration of metals in the soil with and without HouActinite treatment at 0- to 15-, 15- to 30-, and 30- to 45-cm layers of the soil profile; and (iii) changes in cation-exchange capacity and pH with depth of the soil profile.

EXPERIMENTAL

One hundred and eighty-eight hectares of established coastal bermudagrass on Edna (fine, montmorillonitic, thermic Vertic Albaqualf) soil located on the Carl Miller farm of Monaville, TX, were treated with dried sewage sludge (HouActinite) as follows: Fifty-four hectares with no sludge, but enough commercial fertilizer (150–50–5 kg ha^{-1} of nitrogen (N)–phosphorus (P)–potassium (K), according to soil test; 69 ha with 2.8 Mg ha^{-1} HouActinite, and 65 ha with 4.5 Mg ha^{-1} HouActinite. A cyclone manure spreader attached to a tractor was used to dispense the treatment each spring during the period 1989 to 1992.

Forage samples of 1 g were ashed and digested in concentrated HCl. Soil samples of 5 g were extracted with 1 *M* NH_4OAc. The plant digests and soil extracts were analyzed for Zn, Cd, Cu, Pb, and Ni by atomic absorption spectrometry. Cation-exchange capacity (CEC) was determined by the Mehlich (1948) procedure. The experimental design utilized in this study was the method of plot sampling in a completely randomized design as given by Gomez and Gomez (1984).

RESULTS AND DISCUSSION

The mean concentrations of metals found in the forages are given in Table 25–1. Zinc, Cu, and Cd were found to increase significantly with sewage sludge treatment for each harvest during the period 1989 to 1992. Zinc concentration (44–57 mg kg^{-1}) was significantly greater than Cu (5–7 mg kg^{-1}) and Cd (3–4 mg kg^{-1}). The concentration of Pb was found to be ≈2 mg kg^{-1} and the Ni concentration was 0.3 mg kg^{-1} or less, irrespective of treatment. All the concentrations of these metals found in the bermudagrass forages were much less than the accepted ranges permitted by the U.S. Environmental Protection Agency and the U.S. Department of Agriculture.

Table 25–1. Mean concentrations of selected heavy metals in coastal bermudagrass tissue treated with different levels of HouActinite.

Harvest	Zn			Cu			Cd			Pb			Ni		
	1[†]	2[†]	3[†]	1	2	3	1	2	3	1	2	3	1	2	3
	mg kg^{-1}														
								1989							
I	44.04d[‡]	48.18c	51.22b	5.68cd	6.28ab	6.48a	2.92cd	3.18c	4.18a	2.24b	2.38a	2.74a	0.28ab	0.24ab	0.26ab
II	43.64d	47.58c	53.26a	5.48cd	5.50cd	6.46a	3.00c	2.58d	4.02a	2.16b	2.00b	2.02b	0.30a	0.18ab	0.26ab
III	44.18d	48.34c	51.34b	5.30d	5.70cd	5.96bc	3.02c	2.80cd	3.58b	2.08b	2.06b	2.36ab	0.22ab	0.22ab	0.16b
								1990							
I	44.26c	48.14b	51.84a	5.42bc	5.88ab	6.22a	2.68c	3.36b	4.26a	2.04ab	1.86b	1.94ab	0.16a	0.10a	0.08a
II	45.18c	47.94b	51.84a	5.38bc	5.88ab	6.36a	2.70c	3.42b	4.30a	1.90ab	2.08a	1.96ab	0.14a	0.10a	0.14a
III	45.08c	47.18b	52.68a	5.24c	5.90ab	6.28a	2.68c	3.58b	4.14a	1.86b	1.96ab	1.90ab	0.10a	0.10a	0.16a
								1991							
I	44.20d	47.96c	54.28b	5.30d	5.80c	6.06abc	2.76c	3.52b	4.22a	1.96a	1.82a	1.90a	0.12a	0.10a	0.14a
II	44.96d	48.58c	55.08ab	5.14d	5.96bc	6.40ab	2.76c	3.54b	3.98a	1.90a	1.92a	1.88a	0.10a	0.06a	0.08a
III	45.44d	48.02c	55.60a	5.14d	6.10abc	6.44a	2.90c	4.12a	4.24a	1.90a	1.98a	1.90a	0.08a	0.10a	0.08a
								1992							
I	44.94c	48.26c	56.28a	5.20b	6.18a	6.42a	2.70c	3.68b	4.32a	1.88a	2.00a	1.86a	0.12a	0.14a	0.06a
II	44.88c	48.26c	55.96a	5.16b	6.06a	6.40a	2.82c	3.48b	4.42a	1.90a	1.86a	1.90a	0.08a	0.08a	0.06a
III	44.96c	52.52ab	56.56a	5.18b	6.30a	6.46a	3.38b	3.36b	4.34a	1.90a	1.90a	1.90a	0.06a	0.04a	0.10a

[†]Treatment levels of HouActinite (none, 2.8, and 4.5 Mg ha^{-1}).
[‡]Means within columns and rows with the same letters are not significantly different ($P < .05$).

The quantities of the exchangeable form of the metals shown in Table 25–2, except Ni in the soil surface layer, increased significantly with HouActinite treatment as follows: Zn, 6 to 112; Cu, 2 to 6; and Cd and Pb, 0.3 to 2 mg kg^{-1}. In every instance, concentrations of the exchangeable metals decreased significantly with soil depth, and except for Zn the amount found at the 30- to 45-cm soil layer was minute and often nondetected. The latter indicated very little movement of the metals down the profile and little or no possibility for ground water contamination. Exchangeable Ni was quite high in the soil surface layer (2.5–3.8 mg kg^{-1}) and decreased significantly with depth. Soil pH values were always above 6, increasing with HouActinite treatment, and always increased with soil depth irrespective of treatment. The already very low organic matter content and CEC values for this soil showed small (<10%) increases with HouActinite treatment, which were not great enough to be a problem with metal loading. Since the quantities of exchangeable (available) form of metals remained relatively consistent throughout the 4-yr study, it is quite likely that most of the metals added each year in the sludge became unavailable and accumulated in the soil as reported by Mangaroo et al. (1965) in other studies involving radioactive strontium (Sr) and Zn.

In conclusion, an aspect of the role of agriculture in the use of the urban waste, dried sewage sludge, is presented. Where dried sewage sludge is available, it could be used in soils such as the Edna fine sandy loam with suitable monitoring of the heavy metal contents of plant and soil. As shown here the concentration of Zn, Cd, Cu, Pb, and Ni were much lower than the tolerance levels published for soil loading with respect to CEC and do not appear to pose a hazard. Movement of heavy metals down the soil profile was found to be negligible, thereby showing little or no threat to contaminate ground water at such high soil pH.

REFERENCES

Chaney, R.L. 1973. Crop and food chain effects of toxic elements in sludges and effluents. p. 129–141. *In* Recycling municipal sludges and effluents on land. National Association of State Universities and Land-Grant Colleges, Washington, DC.

Day, A.D., and R.K. Thompson. 1986. Fertilizing wheat with dried sludge. Biocycle (Sept.):30–32.

Gomez, K.A., and A.A. Gomez. 1984. Statistical procedures for agricultural research. 2nd ed. John Wiley & Sons, New York.

Greenland, D.J., and M.H.B., Hayes (ed.). 1981. The chemistry of soil processes. John Wiley & Sons. Chichester, UK.

Hinesly, T.D., R.L. Jones, E.L. Ziegler, and J.J. Tyler. 1977. Effects of annual and accumulative applications of sewage sludge on assimilation of zinc and cadmium by corn (*Zea mays*, L.). Environ. Sci. Technol. 11:182–188.

King, L.D., and P.M. Giordano. 1986. Agricultural use of municipal and industrial sludge in the Southern United States. Southern Coop. Series Bul. 314. Dep. of Soil Sci., North Carolina State Univ., Raleigh, NC.

Table 25-2. Mean concentrations of selected heavy metals in soil treated with different levels of HouActinite.

Soil depth	Zn 1[†]	Zn 2[†]	Zn 3[†]	Cu 1	Cu 2	Cu 3	Cd 1	Cd 2	Cd 3	Pb 1	Pb 2	Pb 3	Ni 1	Ni 2	Ni 3
- cm -	mg kg^{-1}														
	1989														
0 – 15	5.46f[‡]	96.42b	111.9a	2.06c	4.04b	5.90a	1.12c	1.52b	2.20a	0.30cd	1.20b	1.36a	3.82a	3.12a	3.42a
15 – 30	4.62f	18.88d	26.84c	1.20cd	1.80cd	2.94bc	0.28d	0.18d	0.34d	0.18d	0.26cd	0.34c	1.38c	1.24c	2.28b
30 – 45	3.54g	10.24e	10.18e	0.10d	0.50cd	0.42cd	0.04e	0.02e	0.02e	0.00e	0.02e	0.00e	0.82cd	0.26d	0.22d
	1990														
0 – 15	9.02e	95.84b	112.4a	1.34e	3.70b	5.84a	0.42c	1.44b	2.18a	0.30c	1.16b	1.40a	2.48c	2.78b	3.18a
15 – 30	6.88f	18.92d	27.70c	1.00f	1.76d	2.98c	0.20d	0.24d	0.26d	0.22c	0.34c	0.32c	1.82d	1.20e	2.34c
30 – 45	3.26g	10.16e	10.18e	0.14h	0.52g	0.36gh	0.04e	0.02e	0.02e	0.00d	0.02d	0.02d	0.90f	0.38g	0.26g
	1991														
0 – 15	8.68f	97.70b	113.0a	1.30e	3.86b	5.76a	0.46c	1.34b	1.96a	0.34b	1.22a	1.34a	2.60b	2.70b	3.32a
15 – 30	7.46g	18.70d	28.52c	0.94f	1.66d	3.12c	0.26d	0.22d	0.28d	0.24b	0.30b	0.32b	1.80d	1.36e	2.30c
30 – 45	3.26h	10.12e	10.12e	0.12h	0.58g	0.40g	0.04e	0.04e	0.02e	0.04c	0.04c	0.04c	0.90f	0.32g	0.30g
	1992														
0 – 15	8.78fg	98.52b	115.9a	1.46d	3.86b	6.34a	0.48b	1.44a	1.64a	0.28b	1.32a	1.38a	2.60b	2.82ab	3.02a
15 – 30	7.82g	19.10d	29.74c	0.94e	1.66d	3.12c	0.26bc	0.32bc	0.24bc	0.20bc	0.30b	0.30b	1.80d	1.58d	2.30c
30 – 45	3.16h	10.40e	9.96ef	0.16g	0.72ef	0.46fg	0.02d	0.04d	0.02d	0.04c	0.04c	0.04c	1.14e	0.32f	0.28f

[†]Treatment levels of HouActinite (none, 2.8 and 4.5 Mg ha^{-1}).

[‡]Means within columns and rows with the same letters are not significantly different (P <.05).

Leeper, G.W. 1978. Managing the heavy metals on the land. Marcel Dekker, New York.

Mangaroo, A.S., F.L. Himes, and E.O. Mclean. 1965. The adsorption of Zn by some soils according to various pre-extraction treatments. Soil Sci. Soc. Am. Proc. 29:242–245.

Mehlich, A. 1948. Determination of cation– and anion–exchange properties of soils. Soil Sci. 66:429–445.

Stevenson, F.J. 1987. Soil biochemistry: Past accomplishments and recent developments. p. 133–143. *In* L.L. Boersma et al. (ed.) Future developments in soil science research. SSSA, Madison, WI.

26 Utilization of Municipal Solid Waste and Sludge Composts in Crop Production Systems

John H. Peverly
P. Brad Gates

Department of Soil, Crop and Atmospheric Sciences
Cornell University
Ithaca, New York

The application of composts to full-scale agricultural crop production systems has not been widely practiced. As pressures build to close solid waste landfills and to recycle basic resources, including organic matter and nutrients, alternative uses are being sought. In addition, composting facilities for a wide variety of inputs are being designed, sited and built.

As a result interest in land application as a beneficial use of municipal solid waste (MSW) and sludge composts is high. Potential benefits include primary plant nutrient fertilization for nitrogen (N) and phosphorus (P) (Mays & Giordano, 1989). In addition, soil rooting characteristics are improved.

Questions still remain, however. The true N and metal mineralization rates for these composts in soil are not well known (Sims, 1990; Woodbury, 1992). Too many guesses are made when calculating N-based application rates, and potential for underfertilization or excessive NO_3 release arises. In fact, the impact of large organic matter additions to soil pore water and ground water quality needs further research from the standpoint of salt and metal contamination also. Finally, the short-term safety from organic chemicals and pathogens must be confirmed by field demonstration and monitoring, as well as the long term fate of metals under recommended management practices.

Therefore, two composts of differing makeup were used in field trials on corn (*Zea mays* L.) to evaluate benefits and risks. This report represents the results of the first 2 yr.

Sewage Sludge: Land Utilization and the Environment. SSSA Miscellaneous Publication.

PROCEDURES

An MSW co-compost (with sludge) from the Delaware Solid Waste Authority and a sludge-plus-woodchip compost produced by the Allgro Corporation were delivered to the application site in Seneca County, New York, in early May each year. Compost analyses for 1993 are presented in Table 26-1, showing a lead (Pb) content for MSW compost near the New York State limit, and a carbon:nitrogen ratio near 10 for both composts. The site was on a Schoharie silty clay loam (fine, illitic, mesic Typic Hapludalf) of slope <3%. The effect of the composts on yield, soil properties, and water quality were assessed and compared with conventionally managed plots, all replicated three times.

Table 26-1. Levels of metals in municipal solid waste (MSW) and sludge composts applied to plots, and upper limits for New York State Class I composts.

	Composts		
Parameter	MSW	Sludge	Class I
	---------------------- mg kg^{-1} ----------------------		
Zinc	800	405	2500
Copper	302	481	1000
Nickel	163	94.1	200
Chromium-total	198	160	1000
Cadmium	3.84	3.12	10
Lead	318	103	250
Mercury	2.7	ND	10
	---------------------- % ----------------------		
Total N	1.46	2.63	-
Organic C	13.7	24.3	-

Composts were applied the third week in May at nominal rates of 46, 92, and 184 Mg ha^{-1} (Table 26-2), the low rates corresponding to agronomic N requirements. The compost was applied to 16 m x 16 m plots (3 replicates), preplant fertilizer broadcast at 26, 105, and 105 kg ha^{-1} for N, P, and potassium (K), respectively and disked in. Corn variety Pioneer 3733 was planted the last week of May and subsequently a normal program of weed and disease control applied by the landowner. The MSW, 92 Mg ha^{-1} plots and control plots were split and fertilizer N as ammonium nitrate applied after planting to attain total rates of 26, 59, 114, and 224 kg ha^{-1} on half the split plots.

Soil samples were collected several times throughout the year by hand corers or auger at depths of 0 to 20 cm and 90 to 100 cm. Nutrient analyses, organic matter and pH determinations were performed by the Cornell University Nutrient Analyses Laboratory. Metals were determined by 8 *M* nitric acid digestion and analysis by inductively coupled plasma (ICP) spectroscopy.

Corn grain was harvested from two in-plot rows for a total of 10.6 m in length and yield adjusted to 15.5% moisture. Plant tissue elemental analysis was based on acidic wet digestion of subsamples of the dried, harvested, ground, plot material. Autoanalyzer techniques were used for P and N, and ICP for other tissue elements.

Table 26-2. Compost treatments and preplant N additions to corn plots, Seneca County, New York, in 1993.

	Treatment[†]				
MSW co-compost	C	DE1	DE2	DE3	CV
Dry wt, Mg ha^{-1}	0	46	92	184	0
Fertilizer N, kg ha^{-1}	26	26	26	26	114
Total available N, kg ha^{-1} (10% mineralization)	26	138	250	474	114
Sludge + woodchip compost	C	MA1	MA2	MA3	CV
Dry wt, Mg ha^{-1}	0	31	62	124	0
Fertilizer N, kg ha^{-1}	26	26	26	26	114
Total available N, kg ha^{-1} (10% mineralization)	26	157	288	550	114

[†]C, compost control; DE, Delaware municipal solid waste (MSW) + sludge; MA, Allgro sludge + woodchips; CV, conventional fertilizer N.

Soil pore water sampling by suction cup lysimeters was conducted at 60 cm on the high rate plots along with controls. Dissolved nutrients and metal elements were analyzed similarly to the extracted soil solutions, with an alternative matrix for the ICP.

Mineralization estimates were obtained by incubating soil samples taken from the compost plots after application and incorporation. Cumulative nitrate production was measured every two weeks after nondestructive leaching of soil samples incubated at 30°C, for a total period of 120 d (Chae & Tabatabai, 1986).

RESULTS AND DISCUSSION

Corn yields in 1992 topped out at about 7.5 Mg ha^{-1} for the high rate sludge compost, close to the potential for this soil. The MSW compost at the highest rate yielded 6.0 Mg ha^{-1}, a reasonable yield for the cool, wet weather in 1992 and the probable loss of N to denitrification and perhaps leaching. Conventional fertilizer treatment yielded 5.4 Mg ha^{-1}. Apparently N was the limiting factor, and there was no indication of any toxic effects of the composts on corn germination or growth in 1992.

Soil pore water samples collected in early spring 1993 (Table 26-3) indicated carryover of mineralized N from compost treatments in the 100 to 200 Mg ha^{-1} range. These pore water NO_3 concentrations, however, were not higher than the controls except for the high rate sludge compost treatment. In addition, soil NO_3 analysis in May 1993 just before compost application showed essentially zero levels. Sodium in pore water was elevated under the high rate MSW compost treatment, but again, no adverse effects on growth or yield were observed from this source in 1992 or 1993. No other metal or nutrient levels above controls were measured in April 1993 pore water samples.

Expected increases in soil pH, organic matter, and total N were documented but are not presented here. Bulk densities decreased consistently from ≈1.5 to 1.2 g cm^{-3} with increasing application rates, but were not significantly different ($P \leq 0.05$).

Five weeks after application in 1993, soil nutrient analyses (Fig. 26-1) showed substantial treatment effects for N, P, and K, with the agronomic N rate (the lowest rate) sludge compost boosting available N up to ≈200 kg ha^{-1}. In fact, the incubation studies of these treated soils indicated a total NO_3 release of ≈350 kg ha^{-1} for this same treatment, while the low rate MSW compost-treated sample showed ≈200 kg ha^{-1} organic N mineralized during incubation (Fig. 26-2). These were substantial quantities in both cases and may indicate over application of N beyond crop needs in the case of corn.

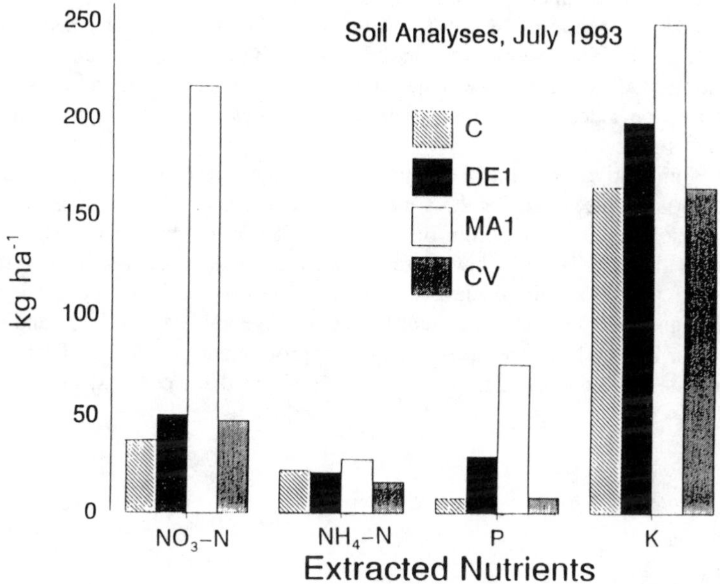

Fig. 26-1. Soil nutrient analysis after the second annual (1993) compost application. C, control; DE1, municipal solid waste compost at 46 Mg ha^{-1}; MA 1, sludge compost at 31 Mg ha^{-1}; CV, conventional fertilizer at 114 kg N ha^{-1}.

Table 26-3. Soil pore water quality at 60 cm under treated plots, in April 1993.

Constituent	Treatment†		
	C	DE3	MA3
	- - Concentration (± standard error) - -		
NO_3–N (mg kg^{-1})	13.7(1.4)	4.37(1.5)	24.4(4.5)
NH_4–N (mg kg^{-1})	0.08(.05)	0.14(.06)	0.11(.05)
Na (mg kg^{-1})	5.52(.46)	18.9(1.52)	6.21(1.48)

†C, control; DE3, 172 Mg ha^{-1} municipal solid waste compost; MA3, 134 Mg ha^{-1} sludge compost. Applied May 1992.

Net mineralization as indicated by the incubation study ranged from a low of 2.9% for low rate MSW compost to a high of 104% for the high rate sludge compost treatment. Unlike the results reported by Sims (1990), there was no net immobilization of N, but N release by MSW compost was substantially less than that by sludge compost (Fig. 26-2). Both the field sample analyses and incubation studies indicate the remaining uncertainty for calculating organic N mineralization rates and compost application rates based on agronomic N requirement.

Cumulative soil metal loading values are presented in Table 26-4. Increases were measured for the six metals listed in all of the compost-treated plots. If soil loading limits from Baker et al. (1985) are used as a standard, zinc (Zn) and nickel (Ni) limits have already been exceeded even for the controls. With corn at these soil pH levels, however, there is probably no cause for concern, and the U.S. Environmental Protection Agency levels at 2800 and 420 kg ha^{-1} for Zn and Ni, respectively, may be appropriate limits to use. There was no consistent yield relationship to soil Ni or Zn at these compost application rates.

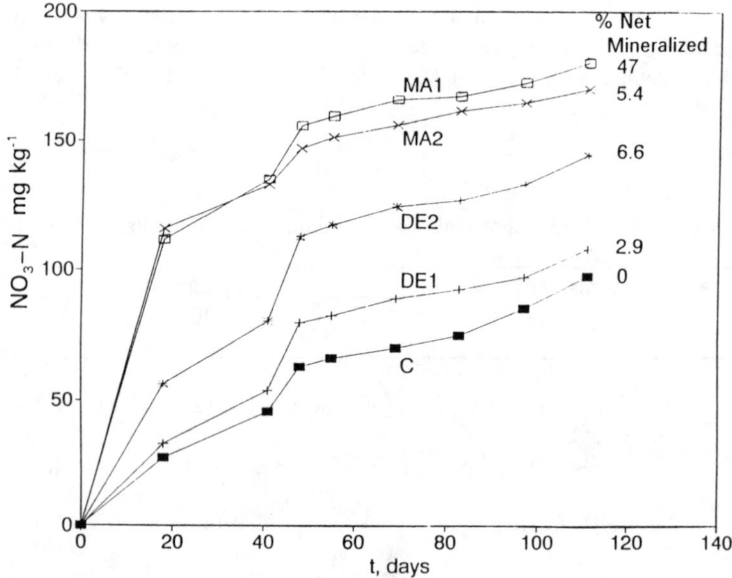

Fig. 26-2. Cumulative organic N mineralization for 110 d of incubation at 30°C. C, control; DE1 and DE2, 46 and 92 Mg ha^{-1} municipal solid waste compost; MA1 and MA2, 31 and 62 Mg ha^{-1} sludge compost.

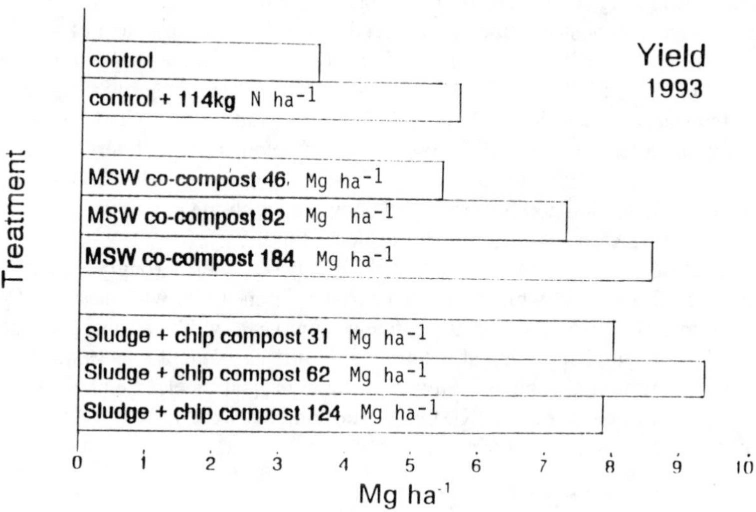

Fig. 26-3. Corn yields in response to second-year compost application and conventional fertilization.

Table 26-4. Soil cumulative load limits for total metals compared with surface soils in 1993 corn plot trials.

Metal	Limits‡	Treatment†			
		C	DE1	MA1	CV
	kg ha^{-1} (± standard error)				
Zn	167	110.0 (10.9)	167.0 (9.0)	14.09 (9.4)	122.0 (15.6)
Cu	82	26.0 (2.0)	60.9 (14.7)	64.8 (3.5)	28.2 (3.4)
Ni	33	36.9 (3.7)	48.4 (4.5)	45.1 (3.2)	39.9 (6.9)
Cr	330	47.8 (5.0)	55.9 (6.8)	59.5 (2.6)	48.7 (8.0)
Cd	3.3	0.88 (0)	1.03 (0.07)	0.97 (0.03)	0.95 (0.07)
Pb	330	47.8 (5.0)	69.8 (5.6)	55.7 (1.1)	50.4 (6.6)

†C, control; DE1, 46 Mg ha^{-1} municipal solid waste compost; MA1, 31 Mg ha^{-1} sludge compost; CV, conventional fertilizer at 114 kg N ha^{-1}.
‡Baker et al. (1985).

Yields for 1993 presented in Fig. 26-3 show expected responses to fertilizer and increasing compost N ranging up to 9.4 Mg ha^{-1}. Low rate MSW and sludge compost treatments produced at least as well as the 114 kg N ha^{-1} conventionally fertilized control yield of 5.6 Mg ha^{-1}. Incubated mineralization values exceeded 200 kg ha^{-1} available N for the low compost rates. So even at the low rates, N should not have been limiting. Indeed, the yield comparisons among the intermediate MSW compost rates supplemented with fertilizer N (Fig. 26-4) also indicate N release rates of >224 kg ha^{-1}. Yields at the highest fertilizer and sludge compost rates fell compared with the next lower application rate, suggesting an inhibitory effect of some kind.

Nutrient and metal content of grain was not strongly affected by treatment (Fig. 26-5), while K, Ni and perhaps Cu in stover was increased by the agronomic N (low compost) rate treatments compared with controls (Table 26-5). The elevated soil Zn did not affect grain or stover Zn content that seemingly was depressed by sludge compost. Stover Ni was significantly elevated above control in response to elevated soil Ni, but stover contents were only a fraction of the recommended limits (Baker et al., 1985).

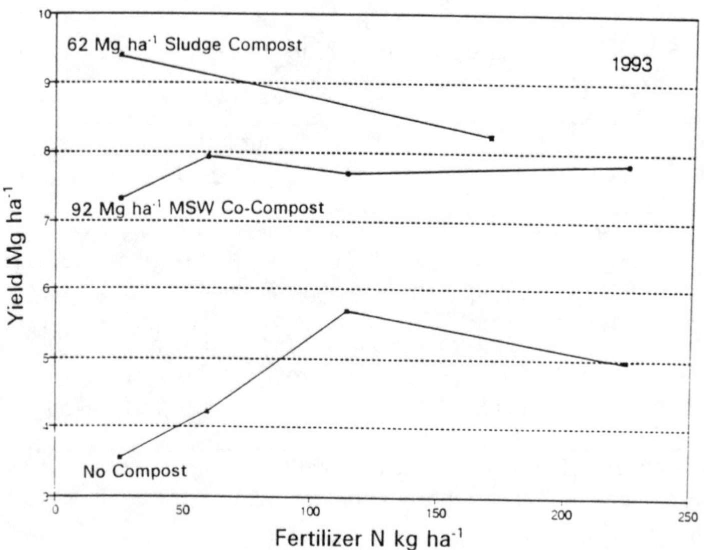

Fig. 26-4. Corn yields in response to second-year compost applications, plus fertilizer N additions as ammonium nitrate.

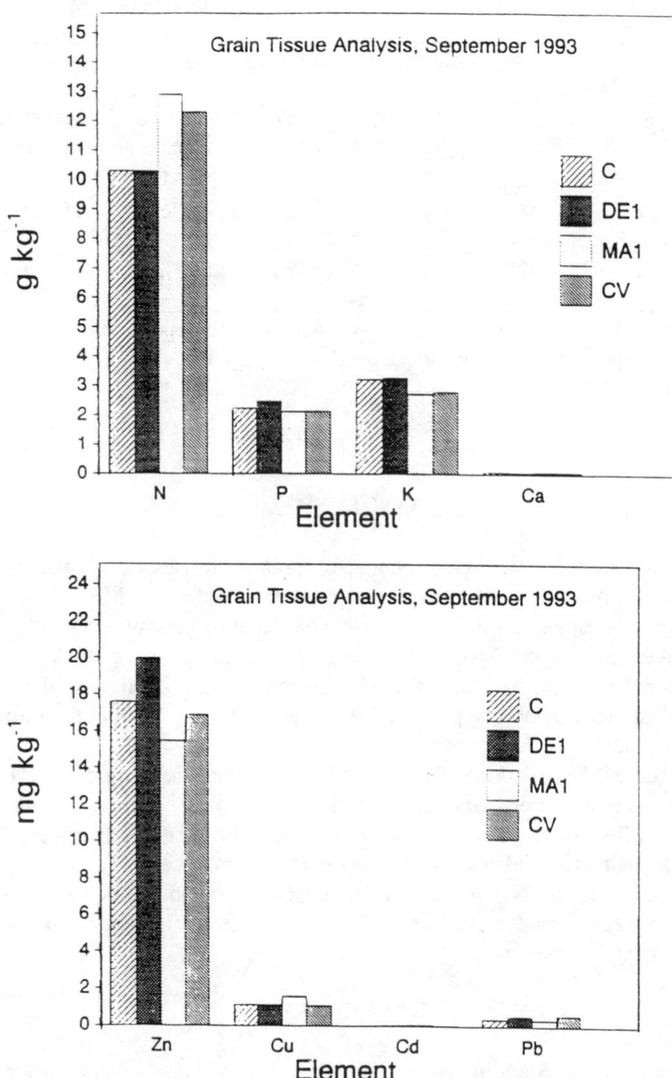

Fig. 26-5. Grain elemental analysis. C, control; DE1, 46 Mg ha^{-1} municipal solid waste compost; MA1, 31 Mg ha^{-1} sludge compost; CV, conventional fertilizer at 114 kg N ha^{-1}.

Table 26-5. Corn tissue limits (silage) for metals compared to 1993 field plot trials.

Metal	Limits‡	Treatment†			
		C	DE1	MA1	CV
		mg kg^{-1} (± standard error)			
Zn	55	21.5 (1.95)	16.6 (2.94)	12.8 (1.74)	24.4 (0.62)
Cu	30	2.0 (0.12)	1.7 (0.20)	4.5 (0.64)	3.7 (0.66)
Ni	3	0.15 (0.03)	0.20 (0.04)	0.21 (0.07)	0.13 (0.04)
Cr	2	0.96 (0.05)	0.86 (0.1)	0.92 (0.07)	0.95 (0.02)
Cd	0.5	0.04 (0.01)	0.07 (0.01)	0.02 (0.01)	0.01 (0.03)
Pb	30	0.73 (0.01)	0.81 (0.08)	0.42 (0.34)	0.71 (0.06)

†C, control; DE1, 46 Mg ha^{-1} municipal solid waste compost; MA1, 31 Mg ha^{-1} sludge compost; CV, conventional fertilizer at 114 kg N ha^{-1}.
‡Baker et al. (1985).

CONCLUSIONS

Short-term, field scale compost applications at agronomic N rates produced no increase in NO_3–N in shallow ground water, but MSW co-compost may cause increases in soil soluble salts (Na), compared with controls.

After two successive years of compost applications, corn yields were again increased compared with those on conventionally fertilized and managed plots. Nitrogen availability probably exceeded 200 kg ha^{-1} even on the low rate compost plots.

The MSW co-compost and sludge-plus-woodchip compost (C/N ≈ 10) applications to a farmer-managed field showed no adverse effects on soil, water, or crop quality, although it is probable that calculated organic N mineralization was underestimated and that N release exceeded crop needs.

Increased soil metals from agronomic compost application rates remained in the top soil. Elevated Zn and Ni soil levels affected neither yield nor grain and water quality.

REFERENCES

Baker, D.E., D.R. Bouldin, H.A. Elliot, and J.R. Miller. 1985. Criteria and recommendations for land application of sludges in the Northeast. North East Regional Res. Publ. Bull. 851. Pennsylvania State Univ., University Park.

Chae, Y.M., and M.A. Tabatabai. 1986. Mineralization of nitrogen in soils amended with organic wastes. J. Environ. Qual. 15:193–198.

Mays, D.A., and P.M. Giordano. 1989. Landspreading municipal waste compost. Biocycle 24:37–39.

Sims, J.T. 1990. Nitrogen mineralization and elemental availability in soils amended with co-composted sewage sludge. J. Environ. Qual. 19:669–675.

Woodbury, P.B. 1992. Trace elements in municipal solid waste composts: A review of potential detrimental effects on plants, soil biota, and water quality. Biomass Bioenergy 3:239–259.

27 Effect of Iron, Manganese, and Zinc Enriched Biosolids Compost on Uptake of Cadmium by Lettuce from Cadmium-Contaminated Soils

R. L. Chaney
C. E. Green

USDA-Agricultural Research Service
Beltsville, Maryland

E. Filcheva

Poushkarov Institute of
Soil Science and Agroecology
Sofia, Bulgaria

S. L. Brown

Department of Agronomy
University of Maryland
College Park, Maryland

The ability of hydrous oxides of iron (Fe) and manganese (Mn) in municipal biosolids to adsorb metals in the biosolid and in amended soils, and limit the phytoavailability of many trace elements has been documented. We conducted an experiment to test whether deliberate addition of Fe or Mn to sludge compost would reduce the potential for cadmium (Cd) uptake by lettuce (*Lactuca sativa* L.) from soils that had been amended with high Cd sewage sludge in the 1970s. Results from this study may be applicable to remediation of contaminated soils, as well as to preparation of *designer biosolids*, which lower metal phytoavailability and bioavailability. Soils were obtained from fields that received high Cd sludge (≈1000 mg Cd kg^{-1} and 1000 mg zinc (Zn) kg^{-1}, dry wt); the Cd was discharged by a user of Cd pigments. In field trials with 2% limestone addition to a depth of 30 cm, lettuce and spinach (*Spinacia oleracea* L.) contained >50 mg Cd kg^{-1}, dry wt. In addition to Fe and Mn addition to increase metal adsorption, Zn addition was included in the study because our research has demonstrated that added Zn inhibits Cd uptake and translocation by plants. The treatments included factorial additions of Fe, Mn, and Zn to the compost; 10% compost by dry weight was added to the test soil; 2% reagent

Sewage Sludge: Land Utilization and the Environment. SSSA Miscellaneous Publication.

$CaCO_3$ was added to all but an acidic control for each soil.

Composted biosolid low in Cd, Zn, Fe, Mn, and $CaCO_3$ was amended with the different combinations and incubated for 1 wk. The treatments were applied to the soils collected from three farmer's fields, incubated for 1 wk, leached to remove soluble salts, fertilized and remixed, and seeded. Romaine lettuce was grown in a growth chamber for five wk, and analyzed for metals of interest.

An unexpected result (Table 27-1) was observed: Zn deficiency was induced by liming the soils. Perhaps the soil Cd interfered with Zn uptake, but plant Zn was well below the 12 to 15 mg kg^{-1} considered diagnostic of deficiency. When lettuce was Zn deficient, greater Cd concentrations were reached. Added Zn or peat + Zn greatly reduced lettuce Cd levels, while

Table 27-1. Effect on lettuce of adding different Zn-, Fe-, and Mn-amended biosolid compost, Zn, peat, or limestone to pH 6.2 soil (historically contaminated by Cd-rich sewage sludge), which contained 6.8 mg Cd and 89 mg Zn kg^{-1}.

Treatments	pH	Yield	Cd	Zn
		g pot^{-1}	---- mg kg^{-1} ----	
Control	5.6	5.9c†	14.3b	12.8e
$CaCO_3$	7.5	1.6d	19.5a	9.9f
$CaCO_3$ + Zn	7.4	6.8abc	1.9g	53.7c
$CaCO_3$ + Peat	7.3	1.2e	12.0bc	9.3f
$CaCO_3$ + Peat + Zn	7.3	7.9ab	2.2f	62.8b
$CaCO_3$ + Compost	7.3	8.4a	9.1de	37.3d
$CaCO_3$ + Compost + Fe	7.1	8.4a	10.1cd	40.3d
$CaCO_3$ + Compost + Mn	7.3	7.8ab	7.7e	35.6d
$CaCO_3$ + Compost + Zn + Mn	7.2	8.4a	7.6e	38.5d
$CaCO_3$ + Compost + Zn	7.2	6.6bc	3.9f	64.7b
$CaCO_3$ + Compost + Zn + Fe	7.1	7.8ab	3.8f	63.1b
$CaCO_3$ + Compost + Zn + Mn	7.3	7.6ab	3.7f	67.8ab
$CaCO_3$ + Compost + Zn + Fe + Mn	7.0	7.0abc	3.9f	75.5a

†Geometric means followed by the same letter are not significantly different ($P \leq 0.05$) according to the Waller-Duncan K-ratio *t*-test.

compost and Zn-amended compost were not quite as effective as the Zn salts. No apparent effects of the added Fe and Mn oxides on lettuce Cd were observed; perhaps the oxides also adsorbed added Zn, reducing its ability to inhibit Cd uptake by lettuce.

We conclude that the application of excess limestone plus added Zn to achieve 100 Zn:1 Cd in the soil will greatly reduce soil Cd risk to feed- and food-chains on these farms with high Cd contamination. This action is achieved through Zn interference with Cd uptake and translocation to lettuce shoots, and by Zn phytotoxicity in soil as pH is allowed to fall. These results should be applicable to many locations where little Zn accompanied Cd contamination of soils.

ACKNOWLEDGMENT

Participation of Dr. Filcheva in this project was supported by USDA-Office of International Cooperation and Development.

28 Sewage Sludge on Acid Mine Spoils: Grasses Produce More Than Legumes

E. Pallant
S. Burke

Department of Environmental Sciences
Allegheny College
Meadville, Pennsylvania

Reclamation of acid mine spoils can be enhanced by applications of lime, sewage sludge, and fast growing grasses and legumes. In a laboratory study three grasses and two legumes were grown for 8 wk on acidic mine spoils treated with dolomitic lime and municipal sewage sludge. On average the grasses produced more than triple the aboveground biomass, more than double the root length density, and approximately double the belowground biomass of the legumes. Two of the grasses had small, but significantly greater leaf N concentrations than the legumes. The legumes, however, had greater root to shoot ratios. We speculated that the legumes we tested had no advantages over the grasses, despite inoculation with rhizobia, because of the abundant nutrient supply in the sludge. Reclaiming mine spoils treated with sewage sludge may be accomplished more effectively using grass species than legumes.

Sewage Sludge: Land Utilization and the Environment. SSSA Miscellaneous Publication.

29 Rangeland Restoration with Treated Municipal Sewage Sludge

Richard Aguilar
Samuel R. Loftin

Rocky Mountain Forest and Range Experiment Station
USDA-Forest Service
Albuquerque, New Mexico

Philip R. Fresquez

Los Alamos National Laboratory
Los Alamos, New Mexico

Municipal sewage sludge could become a valuable resource for restoring the vigor of arid and semiarid rangelands. Before application of sewage sludge in Southwestern rangelands becomes common practice, however, it must be proven both environmentally sound and economically beneficial. A preliminary study during 1985 to 1989, investigating how different quantities of sludge affect vegetative response and plant and soil chemistry, showed that a one-time surface application of municipal sewage sludge at 22.5 to 45 Mg ha^{-1} (10 to 20 ton $acre^{-1}$) significantly increased plant production and ground cover without producing undesirable levels of potentially hazardous elements, including heavy metals, in either soils or plant tissue. A second study evaluated the effects of sludge on rangeland hydrology. Sludge was applied (45 Mg ha^{-1}, surface broadcast) on plots with slope gradients of 6% and 10 to 11%. Runoff from treated plots was immediately curtailed because of increased surface roughness and water absorption by the sludge. An increase in plant cover over time will promote infiltration and further reduce runoff. Surface water contamination by sludge-borne metals and nitrate (NO_3) did not occur during the initial 2 yr of this study. Leaching and translocation of NO_3 through the soil profile did occur on treated plots. Sludge application, however, poses little threat to ground water resources in areas with adequate ground water depths because evapotranspiration potential generally exceeds total precipitation in this semiarid environment.

Southwestern rangelands experienced heavy livestock grazing during the past century, and this led to a substantial reduction in total plant cover and density. Many rangeland soils have been significantly depleted of organic matter and external organic matter additions may be needed to provide sufficient nutrient levels for successful revegetation and establishment of a stable soil organic matter pool.

Sewage Sludge: Land Utilization and the Environment. SSSA Miscellaneous Publication.

The use of municipal sewage sludge for rehabilitating degraded rangeland represents an alternative to the mere disposal of sludge. In the USA, approximately 6 000 000 Mg of municipal sewage sludge are produced annually (U.S. Environmental Protection Agency, 1990). Disposal of this waste product has become a major problem for large metropolitan areas, particularly those along the heavily populated Eastern Seaboard. Large urban areas of the Southwest, including the city of Albuquerque, also produce sewage sludge from wastewater treatment plants and disposal remains a problem. Innovative ways of beneficially using sewage sludge must be continually developed.

Depending upon its source, sludge contains varying concentrations of aluminum (Al), boron (B), cadmium (Cd), copper (Cu), nickel (Ni), lead (Pb), and zinc (Zn) (Sommers, 1977). Ryan et al. (1982) identified Cd as the sludge-borne metal with the greatest potential hazard when sludge is applied to land. Heavy metal movement in soils depends upon pH, cation-exchange capacity (CEC), and organic matter content. Metal mobility, however, is restricted in alkaline arid and semiarid soils due to the formation of insoluble carbonate and phosphate complexes (Bohn et al., 1979). The probability of ground water contamination by NO_3 generated through sludge decomposition and subsequent nitrification is very low in arid and semiarid regions because of low leaching potential.

The benefits of municipal sewage sludge as an organic soil amendment to ameliorate degraded rangeland were tested in two separate studies carried out by the USDA-Forest Service in New Mexico.

SLUDGE APPLICATION RATE STUDY

Field and Laboratory Methods

Dried, anaerobically digested sewage sludge from the City of Albuquerque, NM was surface-applied to a degraded, semiarid grassland site within the Rio Puerco Resource Area ≈100 km northwest of Albuquerque. The Rio Puerco basin, an extremely degraded watershed with a long history of heavy livestock grazing, is one of the most eroded and overgrazed river basins in the arid West (Sheridan, 1981). The sludge was surface-applied (one-time application) at 0, 22.5, 45, and 90 dry Mg ha^{-1} (0, 10, 20, and 40 ton $acre^{-1}$) on each of four plots (3 m x 20 m) in a completely randomized block design containing a total of 16 plots. Minimizing disturbance to the fragile semiarid rangeland site during the sludge application was a major priority. The site was characterized as a broom snakeweed [*Gutierrezia sarothrae* (Pursh) Britton & Rusby]–blue grama [*Bouteloua gracilis* (Willd. ex Kunth) Lagasca ex Giffiths] –galleta [*Hilaria jamesii* (Torrey) Benth.] plant community on a moderately deep, medium-textured soil. Mean annual precipitation, measured at the site with a standard rain gauge throughout the five-growing season study (June 1985 to September 1989) was ≈250 mm (Fresquez et al., 1991).

The study site was fenced to exclude livestock. Pretreatment soil samples were collected from each of the 16 study plots in June 1985. Posttreatment samples were collected in August of each year, 1985 to 1989. Plot sampling scheme, sample preparation and handling techniques, and the chemical characteristics of the applied sludge and the soils prior to the application are described in Fresquez et al. (1991). Methods employed for soil chemical tests and plant tissue analyses for vegetation samples collected from each plot are described in Dennis and Fresquez (1989). Statistical methods employed to test for differences in soil and vegetation properties among the various sludge application treatments are described by Fresquez et al. (1991).

Changes in Soil Nutrients and Heavy Metals

The sludge amendment had a tremendous impact on soil chemistry and nutrient levels. Total nitrogen (TKN), phosphorus (P), and electrical conductivity (EC) increased with sewage sludge application during the study's first year (Table 29-1). Soil organic matter in mineral soil below the sludge layer did not increase appreciably until after the fifth growing season. Fresquez et al. (1991) concluded that the delayed soil organic matter response was an indirect effect of the increased nutrient availability and belowground plant and microbial productivity in response to the sludge amendment.

Soil pH dropped from 7.8 to 7.5 in the 90 Mg ha^{-1} treatment during the first growing season. Soil pH then dropped to 7.4 in both the 45 and 90 Mg ha^{-1} treatments during the second growing season, probably as a result of slightly acidic leachates generated by the applied sludge (Fresquez et al., 1991). Acid-producing microbial reactions in the soil (i.e., nitrification) may have contributed to the decreases in soil pH. Soil pH continued to decrease in plots with the highest sludge application rate as the 5-yr study progressed. Metals generally become more soluble with decreased pH. Despite the decreases in soil pH, however, only diethylenetriaminepentaacetic acid (DTPA)-extractable soil copper (Cu) and cadmium (Cd) increased to concentrations slightly above the limits considered acceptable, and this occurred only after five growing seasons. Concentrations of >10 to 40 mg kg^{-1} Cu and >0.1 to 1.0 mg kg^{-1} Cd are considered phytotoxic and undesirable in the soil (Tiedemann & Lopez, 1982). Changes in other trace elements produced by the different sludge amendments are described in Fresquez et al. (1990b and 1991). The higher trace element concentrations in the sludge-amended soils were probably a direct result of sludge decomposition rather than pre-existing soil micronutrients going into solution (Fresquez et al., 1991).

Table 29-1. Changes in soil chemical properties on plots (n = 4 per application) treated with sewage sludge (Rio Puerco Resource Area, New Mexico; adapted from Fresquez et al., 1991).

Sludge application	Organic matter	TKN†	P	Cd	Cu	EC†	pH
$Mg\ ha^{-1}$	$g\ kg^{-1}$	------ $mg\ kg^{-1}$ ------				$dS\ m^{-1}$	
First growing season							
0	12a‡	729b	5c	0.01a	1.04c	0.36c	7.8a
22.5	13a	817ab	15bc	0.01a	1.19bc	1.06bc	7.7ab
45.0	14a	845ab	20b	0.01a	1.60ab	1.66ab	7.6b
90.0	12a	924a	31a	0.01a	2.10a	2.23a	7.5b
Second growing season							
0	14ab	665b	4c	0.01a	0.92b	0.37b	7.8a
22.5	15a	828ab	20bc	0.01a	2.21ab	0.96ab	7.6ab
45.0	15a	843ab	44b	0.02a	2.99a	1.26ab	7.4b
90.0	12b	987a	72a	0.02a	3.48a	1.97a	7.4b
Fifth growing season							
0	14b	682b	9b	0.01b	0.88b	0.40c	7.8a
22.5	18ab	890b	26ab	0.01b	2.40b	0.66b	7.7a
45.0	26a	1869a	42ab	0.15a	23.52a	0.82ab	7.4b
90.0	23ab	1814a	57a	0.20a	29.78a	0.90a	7.0b

†TKN, total Kjeldahl nitrogen; EC, electrical conductivity.

‡Means within the same column and year followed by the same letter are not significantly different at the 0.05 level by Tukey's multiple range test.

Changes in Blue Grama Forage Production and Quality

Normally, a stimulus such as fertilization increases overall plant production while the diversity of plant species decreases (Houston, 1979; Biondini & Redente, 1986). The sludge amendments at the Rio Puerco plots led to decreases in total plant density, species richness, and species diversity, while blue grama cover and total biomass production significantly increased on treated plots (Fresquez et al., 1990a). The positive effects of the sludge amendments on forage production are demonstrated by changes in blue grama production after the first, second, and fifth growing seasons (Table 29-2). Blue grama production was significantly greater for all of the sludge amendments during the first and second growing seasons, with yields ranging from 1.5 to almost 3.0 times that in the nonamended (control) plots.

Table 29-2. Blue grama production (mean production and standard error (SE); $n = 4$) in control and sludge-amended plots after one, two, and five growing seasons (Rio Puerco Resource Area, New Mexico).

Treatment	Production	SE
Mg ha^{-1}	kg ha^{-1}	
First growing season, 1985	(precipitation = 147 mm)	
0	270b[†]	22
22.5	480ab	96
45.0	433ab	100
90.0	509a	62
Second growing season, 1986	(precipitation = 239 mm)	
0	392b	76
22.5	575ab	163
45.0	824ab	114
90.0	1067a	227
Fifth growing season, 1989	(precipitation = 201 mm)	
0	281a	39
22.5	291a	75
45.0[‡]	506a	51
90.0	500a	178

[†]Means within the same column and year followed by the same letter are not significantly different at the 0.05 level by Tukey's multiple range test.
[‡]Significantly different from the control at the 0.10 level by Dunnett's multiple comparison test.

Summer precipitation during 1986 was exceptionally high and the highest yields of dry matter production occurred during this growing season. Blue grama production remained higher in the 45 and 90 Mg ha^{-1} sludge-amended plots after the fifth growing season, although the benefits of the added sludge had greatly diminished for the lowest (22.5 Mg ha^{-1}) sludge amendment. Blue grama production for the 45 and 90 Mg ha^{-1} amendments remained nearly double that of the controls during the fifth growing season; but within-treatment variation also increased, resulting in statistically non-significant differences ($\alpha = 0.05$) between the control and sludge-amended plots.

The sludge amendments also led to significantly increased nutritional value of blue grama tissue. Tissue N, P, K, and crude protein in the blue grama increased with the application of sludge (Fresquez et al., 1991). Furthermore, most of the trace metals, including Cu and Cd, in blue grama plant tissue did not increase significantly during the 5-yr study, thereby alleviating concerns that these toxic elements might be transferred to grazing animals. This is a significant finding because concerns over heavy metal accumulations frequently limit sewage sludge application to land. Based on these cumulative results, Fresquez et al. (1991) concluded that a one-time sludge treatment ranging from 22.5 to 45 Mg ha^{-1} (10 to 20 ton acre^{-1}) produced the best vegetation response in semiarid rangeland without potential harm to the environment.

SURFACE HYDROLOGY STUDY

A second study was established in spring 1991 within the Sevilleta National Wildlife Refuge, ≈120 km south of Albuquerque. The objectives of this study were to determine if and how changes in vegetation following sludge application influence runoff and surface water quality and to assess the fate of potential sludge-borne contaminants introduced to the environment through the addition of sludge. The Sevilleta refuge, managed by the U.S. Department of Interior-Fish and Wildlife Service, was an excellent location to study rangeland treatment effects because public access is restricted and livestock grazing is prohibited. Climate at the refuge area is arid to semiarid with mean annual precipitation ranging from 200 to 250 mm. Within the study area, a blue grama/hairy grama (*B. hirsuta Lagasca*) dominated community was selected for study on a moderately sloping (6%) and strongly sloping component (10 to 11%) of a stable alluvial fan. The deep, well drained Harvey soil was characterized as a fine-loamy, mixed, mesic Ustollic Calciorthid formed in local alluvium and colluvium derived from sandstone and limestone.

Field and Laboratory Methods

Six pairs of runoff plots, each pair consisting of a treated (sludge-amended) and a control (no sludge) plot were established within two hillslope gradient classes (three treated–control plot pairs per slope gradient class). Runoff plot dimensions (3 x 10 m) were identical to those used by USDA-Agricultural Research Service (1987) investigators involving the Water Erosion Prediction Project (WEPP). Therefore, results from this study might be applied to WEPP models for larger-scale predictions on runoff and sediment yields from semiarid grassland ecosystems. The experimental plots were bordered by metal flashing to prevent external water from entering the plots. The borders directed internal surface runoff to the base of the plots during rainfall events, wherein the runoff water was collected in sample reservoirs (Aguilar et al., 1994). A one-time surface application of 45 Mg ha^{-1} of municipal sewage sludge (dry-weight basis), provided by the Albuquerque Public Works Department, was applied to the plots in April 1991.

Total precipitation at the site was measured with two standard rain gauges (rainfall collection buckets) and a self-activating recording rain gauge that recorded storm intensity (mm h^{-1}) as well as total precipitation. The runoff plots were subjected to simulated rainfall in September 1991 after the vegetation had an entire growing season to respond to the sludge treatment, and then again in September 1992. The simulated rainfall was equivalent to a high intensity summer thunderstorm common in the region (4 to 8 cm h^{-1} for 30 min) and the rainfall was distributed simultaneously to each plot pair so differences in infiltration and runoff yield between control and treated plots could be monitored and recorded. Representative samples of the runoff water were collected after each rainfall event and then analyzed for NO_3–N and trace element concentrations. Pre-treatment soil and vegetation characterization established uniformity between control plots and those subsequently treated with sludge. Analytical tests for the soils, vegetation, and the runoff water followed standard procedures as outlined in Page et al. (1982) and Richards (1969). Analysis of variance techniques were used to test for significant differences between the treated and control plots, runoff yield, and runoff water quality.

Hydrologic Response to the Sludge Amendment

First-year natural storm runoff yields were significantly lower from sludge-amended plots than from control plots. Runoff from control plots was 3.4 to 37 times greater than runoff from treated plots (Aguilar et al., 1994). Natural storms did not produce measurable runoff at the study site in 1992 from either the control or treated plots.

In September 1991 and 1992, rainfall simulation experiments were conducted on the runoff plots (Fig. 29-1). Runoff quantities from control plots during these experiments exceeded runoff from treated plots by 27 to >250 times. Runoff yields from our control plots were comparable to runoff yields measured during studies conducted in rangeland elsewhere in New Mexico and Arizona (Ward & Bolton, 1991). Therefore, the hydrologic differences observed between the treated and control plots can be directly attributed to the sludge treatment. The two factors we considered most important for the reduction in runoff on treated plots were increased ground surface roughness and water absorption by the dry sludge. Through time, the sludge should decompose and less directly influence surface runoff processes. Future increases in plant productivity and ground cover, however, could sustain reduced runoff yields from the treated plots.

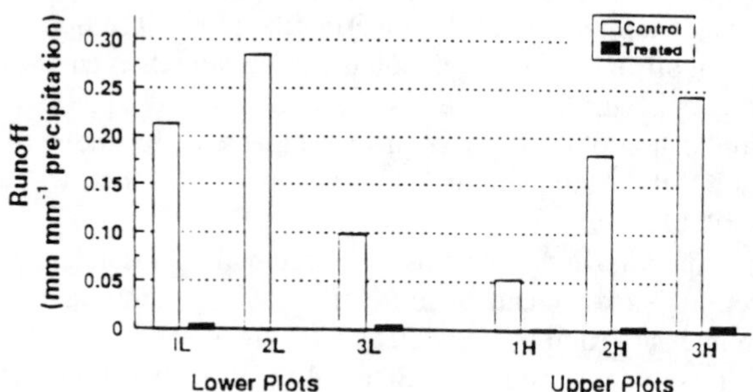

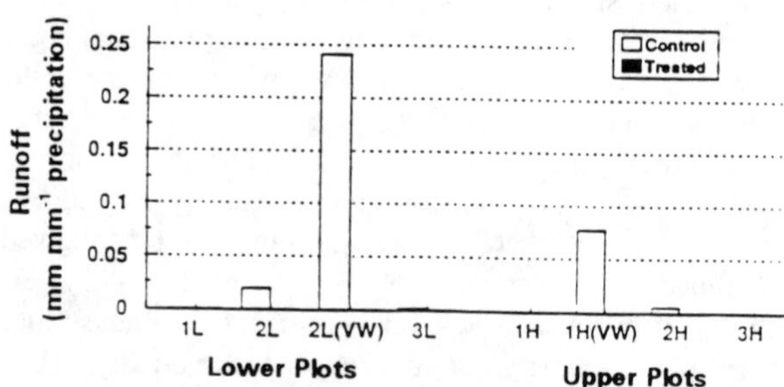

Fig. 29-1. Runoff from sludge-amended (treated) and unamended (control) plots during rainfall simulation experiments. Expression of runoff yield as runoff per mm of precipitation standardizes the runoff for comparison among plots because precipitation input among and between the plot pairs varied somewhat due to wind gusting.

The importance of antecedent soil water on infiltration and runoff was demonstrated during our second series of rainfall simulation experiments, September 21 to 24, 1992 (Fig. 29-1b). Total precipitation from natural storms between May 15 and September 21, 1992 at the study site was only 97.5 mm compared to 182 mm for the same time interval the previous year (1991). Therefore, soil moisture conditions prior to the 1992 simulated rainfall experiments were very low. Because of these dry soil conditions, infiltration rates readily exceeded precipitation inputs, and little runoff occurred from the plots (control or treated) during the initial 1992 rainfall simulation runs (dry runs). Runoff was generated from the control plots only when the initial dry runs were followed with a second 40 to 50 mm rainfall application (very wet run - VW) 15 min later. The very wet runs, however, still did not generate runoff on the treated plots despite cumulative rainfall inputs (dry runs + very wet runs) ranging from 70 to 109 mm.

Potential contamination of surface water by sludge-borne constituents in Albuquerque sludge does not appear to be a limitation for application on degraded rangeland. Nitrate–N, Cu, and Cd concentrations in runoff water were well below New Mexico ground water and livestock and wildlife watering limits, both during natural and simulated rainfall and no statistical differences (α = 0.05) in these potentially toxic constituents were found between the treated and control plots (Aguilar et al., 1994).

CONCLUSIONS

Surface application of treated municipal sewage sludge can significantly increase vegetation cover and total forage production and reduce runoff in semiarid rangeland. Increased ground surface roughness and increased soil water-holding capacity reduce the rangeland's potential for runoff and water erosion. Subsequent increases in vegetation cover due to the sludge's fertilizer effect should further improve the surface hydrology of treated areas. Potential degradation of surface water by sludge-borne contaminants, including heavy metals, in Albuquerque sewage sludge does not appear to be a problem with a one-time application of 22.5 to 45 Mg ha^{-1} (10 to 20 ton $acre^{-1}$). Similar results could be expected using comparable sludge from other municipalities.

ACKNOWLEDGMENTS

Portions of this research were conducted in cooperation with the USDI Bureau of Land Management. The hydrology study was funded by the 1991 New Mexico Water Resources Research Institute (WRRI) - Chino Mines Company Grant Fund and the 1992 WRRI General Grant Program. We would like to thank Dr. Timothy Ward and Dr. Susan Bolton, Dep. of Civil, Agricultural and Geologic Engineering, New Mexico State University, for their assistance with the rainfall simulation experiments. We also thank the City of Albuquerque for providing the sludge and transporting it to our study sites.

REFERENCES

Aguilar, R., S.R. Loftin, T.J. Ward, K.A. Stevens, and J.R. Goez. 1994. Sewage sludge application in semiarid grasslands: Effects on vegetation and water quality. New Mexico Water Resources Research Institute (WRRI) Techn. Completion Rep. 285. New Mexico WRRI, New Mexico State Univ., Las Cruces.

Biondini, M.E., and E.F. Redente. 1986. Interactive effect of stimulus and stress on plant community diversity in reclaimed lands. Reclam. Reveg. Res. 4:211–222.

Bohn, H.L., B.L. McNeal, and G.A. O'Connor. 1979. Soil chemistry. John Wiley & Sons. New York.

Dennis, G.L., and P.R. Fresquez. 1989. The soil microbial community in a sewage sludge amended semiarid grassland. Biol. Fert. Soils 7:310–317.

Fresquez, P.R., R. Aguilar, R.E. Francis, and E.F. Aldon. 1991. Heavy metal uptake by blue grama growing in a degraded semiarid soil amended with sewage sludge. J. Water Air Soil Pollut. 57-58:903–912.

Fresquez, P.R., R.E. Francis, and G.L. Dennis. 1990a. Soil and vegetation responses to sewage sludge on a degraded semiarid broom snakeweed/ blue grama plant community. J. Range Manage. 43:325–331.

Fresquez, P.R., R.E. Francis, and G.L. Dennis. 1990b. Effects of sewage sludge on soil and plant quality in a degraded semiarid grassland. J. Environ. Qual. 19:324–329.

Houston, J. 1979. A general hypothesis of species diversity. Am. Nat. 113:81–101.

Page, A.L., R.H. Miller, and D.R. Keeney. (ed.) 1982. Methods of soil analysis, Part 2. 2nd ed. Agron. Monogr. No. 9. ASA and SSSA, Madison, WI.

Richards, L.A. (ed.) 1969. Diagnosis and improvement of saline and alkali soils. USDA Agric. Handb. 60, U.S. Department of Agriculture, Washington, DC.

Ryan J.A., H.R. Pahren, and J.B. Lucas. 1982. Controlling cadmium in the human food chain: A review and rationale based on health effects. Environ. Res. 28:251–302.

Sheridan, D. 1981. Desertification of the United States. Council on Environmental Quality. U.S. Gov. Print. Office, Washington, DC.

Sommers, L.E. 1977. Chemical composition of sewage sludge and analysis of their potential use as fertilizers. J. Environ. Qual. 6:225–232.

Tiedemann, A.R., and C.F. Lopez. 1982. Soil nutrient assessments of mine spoils. p. 66–79. *In* E.F. Aldon, and W.R. Oaks (ed.) Reclamation of mined lands in the Southwest. Soil Conservation Society of America, New Mexico Chapter, Albuquerque, NM.

U.S. Department of Agriculture, Agricultural Research Service. 1987. User requirements USDA-Water Erosion Prediction Project (WEPP) Rep. 1. USDA-ARS, Nat. Soil Erosion Res. Lab., West Lafayette, IN.

U.S. Environmental Protection Agency. 1990. 40 CFR - Part 503, National sewage sludge survey: Availability of information and data and anticipated impacts on proposed regulations. Federal Register 55(218):47210–47283. U.S. Gov. Print. Office, Washington, DC.

Ward, T.J., and S.M. Bolton. 1991. Hydrologic parameters for selected soils in Arizona and New Mexico as determined by rainfall simulation. New Mexico Water Resources Research Institute (WRRI) Techn. Completion Rep. 259. New Mexico WRRI, New Mexico State Univ., Las Cruces.

30 Semiarid Rangeland Response to Municipal Sewage Sludge: Plant Growth and Litter Decomposition

Samuel R. Loftin
Richard Aguilar

Rocky Mountain Forest and Range Experiment Station
USDA–Forest Service
Albuquerque, New Mexico

Municipal sewage sludge was used as a soil organic matter and plant nutrient amendment in an attempt to improve semiarid rangeland conditions. Albuquerque sewage sludge was surface-applied (45 dry Mg ha^{-1}) to plots established on the Sevilleta National Wildlife Refuge in central New Mexico. Sewage sludge application significantly increased above-ground plant cover when water was not limiting. Sewage sludge effects on plant litter decomposition, an important process in nutrient cycling, were inconclusive. While we found no significant sludge treatment effects on decomposition of saltbush [*Atriplex confertifolia* (Torrey & Fremont) Watson] tissue during either of two field seasons, we did observe significant treatment effects on the decomposition of blue grama [*Bouteloua gracilis* (Willd. ex Kunth) Lagasca ex Griffiths] tissue in one of the two field seasons. These results suggest that plant litter decomposition on sludge-amended rangeland is controlled by complex interactions between edaphic and climatic variables and plant litter quality and that these interactions may override the sludge treatment effects.

Relatively little work has been conducted on reclamation of arid rangelands with organic matter amendments. Results from mulching experiments using bark and straw showed some short-term benefits to the soil microbial community, but little improvement in aboveground plant cover (Fresquez & Lindemann, 1982; Whitford et al., 1989). A study conducted in the Rio Puerco Resource Area in northwestern New Mexico tested rangeland vegetation response to surface-applied sewage sludge applications of 22.5, 45 and 90 Mg ha^{-1} (Fresquez et al., 1990). Aboveground plant cover on all treated plots increased significantly over the controls within months following the sludge application. The intermediate application of 45 Mg ha^{-1} was recommended to safeguard against potential heavy metal contamination.

Sewage Sludge: Land Utilization and the Environment. SSSA Miscellaneous Publication.

MATERIALS AND METHODS

The study was conducted within the Sevilleta National Wildlife Refuge (SNWR) in central New Mexico. Climate at the SNWR is arid to semiarid with mean annual precipitation ranging from 200 to 250 mm. Summers are relatively hot and winters are cool. Vegetation on the refuge is dominated by semiarid shortgrass prairie and great basin shrubland at low elevations and pinon (*Pinus edulis* Engelm.)–juniper (*Juniperus communis* L.) woodland at higher elevations.

A blue grama/hairy grama dominated community was selected for study on moderately sloping and strongly sloping components of a stable alluvial fan. The typical vegetation assemblage at the study site includes the grasses blue grama, hairy grama (*B. hirsuta* Lagasca), black grama [*B. eriopoda* (Torrey)], ring muhly [*Muhlenbergia torrei* (Kunth) Hitchc.], three awn (*Aristida spp.*), the forb yellow spiny-aster [*Haplopappus spinulosus* (Pursh) DC.], and the shrubs narrow-leaf yucca (*Yucca glauca* Nutt.), winterfat [*Ceratoides lanata* (Pursh) J.T. Howell], four-wing saltbush and groundsel (*Senecio douglasii* DC. var. *longilobus* L. Bensen). Soils at the site are mapped as the Harvey-Dean Association (U.S. Department of Agriculture-Soil Conservation Service, 1988). The Harvey soil is classified as fine-loamy, mixed, mesic Ustollic Calciorthid and the Dean soil is classified as fine-loamy, carbonatic, mesic Ustollic Calciorthid. The deep and well drained soils are formed from alluvium and colluvium derived predominantly from sandstone and limestone.

An important factor determining the response of grassland vegetation to sewage sludge is the distribution of rainfall following sludge application. Roughly half of this precipitation falls during the cool season and the other half occurs during the summer monsoon season. The rainfall measurements for this project were pooled into three, 6-mo periods (Summer-Fall 1991, Winter-Spring 1992, and Summer-Fall 1992). These summer-fall and winter-spring periods correspond with the warm (monsoon) and cool growing seasons, respectively, of the central Rio Grande Valley in New Mexico.

The sludge treatment consisted of a one-time surface application of municipal sewage sludge (45 Mg ha^{-1} or 20 ton acre^{-1} on an oven-dry weight basis) provided by the City of Albuquerque. The sludge was surface-applied, rather than incorporated into the soil, in order to minimize disturbance of the soil and the existing native vegetation. The sludge was applied in early April 1991 to minimize losses that might have occurred during high intensity summer and fall convective storms.

Aboveground plant cover estimates were taken from 10 x 10 m sampling plots (five control and five sludge-amended). These plots were also used for experiments designed to test the effects of sludge application on aboveground and belowground litter decomposition. Changes in aboveground plant cover were determined using a line-intercept technique. Five, 10-m transects were used to obtain estimates of vegetation abundance within each plot.

Treatment effects on aboveground and belowground plant litter decomposition were evaluated on both blue grama and four-wing saltbush tissue. Ten bags of each litter type were placed on the ground surface in each plot and 10 additional bags were buried 5 cm belowground. The litter bags were left in the field for one complete growing season (spring through fall). Samples, within a plot, from the same litter type–position combination were combined, resulting in a total of four samples per plot (blue grama aboveground and belowground and saltbush aboveground and belowground samples). The samples were ground and ashed to correct for mineral content. These procedures were repeated during the 1991 and 1992 growing seasons.

Plant litter samples were oven dried at 80°C for 48 h and ground prior to analyses. The samples were analyzed for total nitrogen (N) content using the total Kjeldahl nitrogen (TKN) method (Bremner & Mulvaney, 1982). Percentage carbon (OC) was estimated as ash-free dry weight/2 and percentage of water soluble constituents were estimated by leaching in tapwater.

Analysis of variance (ANOVA) techniques were used to test for statistically significant differences between treated and control plots in plant cover and plant litter decomposition. Unless otherwise noted, a Type I error rate of $\alpha = 0.05$ was adopted for all of our statistical analyses.

RESULTS AND DISCUSSION

Aboveground Plant Cover

Prior to sludge application (pretreatment sampling period), no significant differences in plant cover were observed between the control and treated plots (Fig. 30-1). One growing season following the sludge application (Fall 1991), plant cover on treated plots was less than but not significantly different from control plots. By Spring 1992, the aboveground plant cover on treated plots had increased to levels greater than, but not significantly different from, control plots. The treated plot plant cover in Spring 1992, however, was significantly greater than Fall 1991 treated plot plant cover. Aboveground plant cover on treated plots then decreased significantly from Spring 1992 to Fall 1992 to levels less than, but not significantly different, from control plots.

Wight and Black (1979) found that the quantity and periodicity of rainfall following a fertilizer application controlled vegetative response. Precipitation during the 1991 monsoon season was >180 mm, however, no significant increase in aboveground plant cover was observed. The delayed plant cover response was partially due to the application of wet sludge on the soil surface, which covered much of the existing vegetation. Vegetation then had to grow out from under the sludge before it could be accounted for using the line intercept transect technique. Even though rainfall was above average during this period, sludge-induced increases in plant cover did not exceed the initial cover loss incurred from the sludge application. The 1991–1992 Winter–Spring period received ≈200 mm of rainfall. This relatively moist period was followed by a comparably dry 1992 monsoon period (≈89 mm).

Treated plot aboveground plant cover response in Spring 1992 resulted primarily from an increase in cool season forbs and shrubs. This growth period was followed by a decrease in forbs and shrubs by Fall 1992. The grasses, mostly warm season plants, could not benefit from the moist, but cool, conditions during the Winter–Spring period and then did not receive adequate

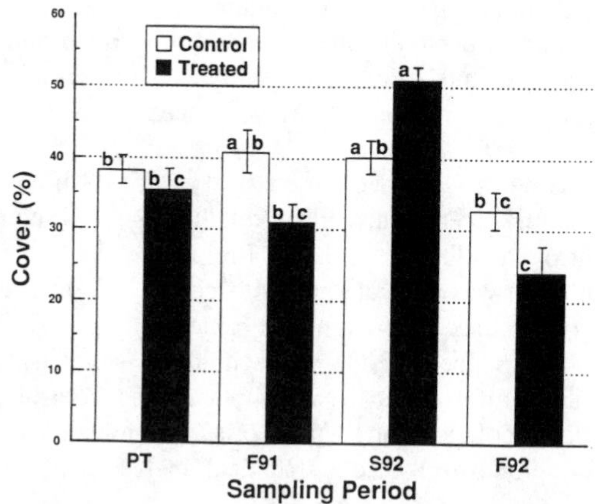

Fig. 30-1. Total aboveground plant cover on control and sludge-treated plots. Data presented are from four sampling periods: pretreatment (PT), Fall 1991 (F91), Spring 1992 (S92), and Fall 1992 (F92). Values reported are means (n = 5) with standard errors. Bars with the same letter are not significantly different at the $P \leq 0.05$ level.

water during the following warm growing season.

The seasonal response in plant cover raises questions related to the plant community dynamics at the study site. It is possible that interactions between the applied sludge and local climatic factors could alter the composition of the normal vegetation assemblage. Many of the cool season species responsible for the overall increase in plant cover in Spring 1992, such as yellow spiny-aster and snakeweed, exhibit the C_3 photosynthetic pathway. The majority of warm-season species, particularly the grasses, exhibit the C_4 photosynthetic pathway. Rauzi et al. (1968) and Smoliak (1965), in separate rangeland fertilization studies, also observed that timing of precipitation and increased N availability produced an increase in cool season (C_3) plants or a decrease in warm season (C_4) plants.

Studies have shown that fertilization of blue grama grassland can increase grass biomass production (Dwyer, 1971; Power & Alessi, 1971; Pieper et al., 1973) and aboveground grass cover relative to shrubs (Pieper et al., 1973). Sewage sludge application to blue grama rangeland was also shown to increase grass production and cover and decrease shrub abundance (Fresquez et al., 1990). Other studies have shown, however, that the response of blue grama to fertilization can be inhibited by unusually low precipitation (Rauzi et al., 1968) or competition with other grasses (Smoliak, 1965). In our study, soil N availability remained high, through Fall 1992, and thus there is still a potential for future increased grass production if water availability does not become limiting.

Plant Tissue Decomposition

Accumulation of soil organic matter and organically-bound plant nutrients represents an important step in the recovery of degraded lands. Decomposition of organic matter and subsequent mineralization of associated plant nutrients is necessary for the long-term growth, productivity, and stability of a restored ecosystem. The sludge treatment effect on the surface (aboveground) and buried (belowground) decomposition of two plant tissue (litter) types was evaluated within control and treated plots.

Figure 30-2 shows the results of the decomposition studies for blue grama grass (Bogr) litter. During the 1991 growing season, significantly greater weight loss occurred in the buried litter bags than in the surface bags, but there were no significant treatment effects. Both litter location (aboveground or belowground) and treatment were significant during the second growing season. Decomposition of saltbush (Atca) litter was significantly greater belowground than at the ground surface during both growing seasons (Fig. 30-2). We found no significant sludge treatment effects, however, in either of the two growing seasons.

The most intriguing questions arising from these decomposition studies is why Bogr litter experienced significant weight loss on sludge-treated plots in 1992, but not in 1991, and why we did not observe the same phenomenon for the Atca litter. As was previously stated, climatic conditions were noticeably different between the two growing seasons. The first growing season following the sludge application was relatively moist (180 mm), while the second growing season was comparably dry (75–80 mm). Additionally, soil N and P levels differed from one season to the next. Soil ammonium was relatively high, and NO_3 and PO_4 were relatively low, at the end of the 1991 season in comparison to the levels at the end of the 1992 season. Interactions among the aforementioned factors, coupled with the initial differences in plant litter characteristics, make interpretation of the decomposition study data difficult. There were no significant initial differences in organic C content between the two litter types, however, Atca litter had a significantly lower C/N ratio and a higher percentage of water soluble constituents (Table 30-1). Apparent differences in weight loss between the two tissue types may have resulted from losses in soluble salts in Atca litter, rather than actual differences in decomposition between the two plant litter types. Because we did not observe similar decomposition losses with both litter types, we conclude that complex interactions among soil and plant tissue variables, and not climate, were chiefly responsible for the differences in litter decomposition.

While our study has not clearly demonstrated sludge treatment effects on plant litter decomposition, other studies have documented an increase in microbial activity following sludge application to soils (Stevenson et al., 1984; Seaker & Sopper, 1988; Dennis & Fresquez, 1989). The apparent discrepancy between our study results and the results of other research may be due to several factors. Other studies did not differentiate between microbial activity associated with the decomposition of sludge organic matter and the decomposition of plant litter, and a large percentage of sludge organic matter is readily decomposed

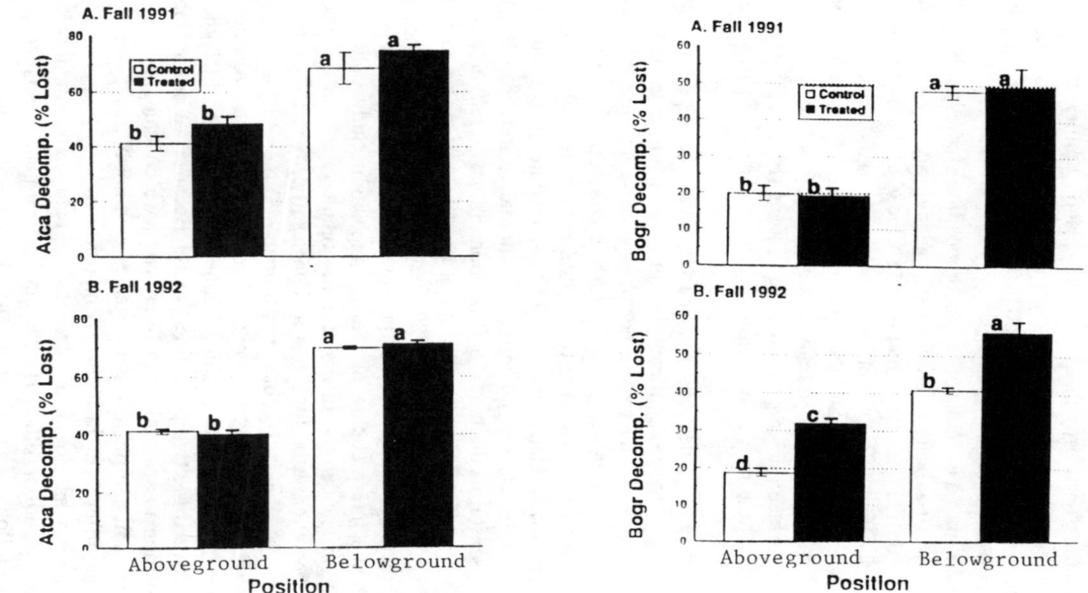

Fig. 30-2. Blue grama (Bogr) and Four-wing saltbush (Atca) litter decomposition on control and sludge-treated plots. Data presented are the results of two separate decomposition experiments conducted during the 1991 and 1992 growing seasons (Spring–Fall). Values reported are means (n = 5) with standard errors. Bars with the same letter, within a season, are not significantly different at the $P \leq 0.05$ level.

Table 30-1. 1992 blue grama (Bogr) and saltbush (Atca) litter quality analysis.

Variable	Bogr	Atca
OC (%)	45.0a[†]	41.0a
TKN (%)	0.27b	1.33a
C/N ratio	167a	31b
Water soluble constituents (%)	3.9b	22.7a

[†]Means followed by the same letter, within the same row (same variable), are not significantly different at the $P \leq 0.05$ level.

(Hartenstein,1981). It is therefore possible that increased microbial activity could be associated with sludge-borne C sources rather than residual soil C pools. It is also possible that the soil N limitations on microbial activity, prior to sludge application, in the other studies were greater than the soil N limitations in our study.

CONCLUSIONS

Surface application of municipal sewage sludge can significantly increase aboveground plant cover in semiarid rangeland. Climatic variables including temperature, and the timing, quantity, and intensity of rainfall are important in determining the rate of recovery and the species composition of the responding vegetation. A moist Winter–Spring period (1991–1992) produced an increase in cool season forb and shrub cover with little response from warm season grasses.

Organic matter decomposition releases organically bound nutrients and makes them available for assimilation by plants and soil microorganisms. This mineralization–immobilization process is necessary for the establishment of a healthy, self-sustaining ecosystem. Sewage sludge application had no detrimental effect on the decomposition of tissues from two plant species, saltbush and blue grama at our study sites. Decomposition of saltbush tissue was not significantly affected by the sludge treatment in either of the two field seasons (1991 and 1992), while decomposition of blue grama tissues was significantly greater on sludge-treated plots only during the 1992 field season. The differences in decomposition between the two tissue types (species) were attributed to interactions among tissue quality (C, N, and soluble salt content), and environmental conditions (soil water and soil nutrient availability).

Although our conclusions are based on only two growing seasons and are therefore preliminary, we believe that surface application of sewage sludge will prove to be a useful tool in the reclamation and management of semiarid rangelands.

ACKNOWLEDGMENTS

This study was funded in part by the 1991 New Mexico Water Resources Research Institute (WRRI) - Chino Mines Company Grant Fund and the 1992 WRRI General Grant Program. We thank the City of Albuquerque, NM for providing the sludge and transporting it to our study sites.

REFERENCES

Bremner, J.M., and C.S. Mulvaney. 1982. Nitrogen–Total. p. 595–624. *In* A.L. Page et al. (ed.) Methods of soil analysis. Part 2. 2nd ed. Agron. Monogr. 9. ASA and SSSA, Madison, WI.

Coker, E.G. 1983. The use of sewage sludge in agriculture. Water Sci. Techn. 15:195–208.

Dennis, G.L., and P.R. Fresquez. 1989. The soil microbial community in a sewage-amended semi-arid grassland. Biol. Fertil. Soils 7:310–317.

Dwyer, D.D. 1971. Nitrogen fertilization of blue grama range in the foothills of south-central New Mexico. New Mexico State University Agric. Exp. Stn. Bull. 585. Las Cruces, NM.

Fresquez, P.R., R.E. Francis, and G.L. Dennis. 1990. Soil and vegetation responses to sewage sludge on a degraded semiarid broom snakeweed/blue grama plant community. J. Range Manage. 43:325–331.

Fresquez, P.R., and W.C. Lindemann. 1982. Soil and rhizosphere microorganisms in amended mine spoils. Soil Sci. Soc. Am. J. 46:751–755.

Hartenstein, R. 1981. Sludge decomposition and stabilization. Science (Washington, DC.) 212:743–749.

Pieper, R.D., D.D. Dwyer, and W.W. Wile. 1973. Burning and fertilizing blue grama range in south-central New Mexico. New Mexico State University Agric. Exp. Stn. Bull. 611. Las Cruces, NM.

Power, J.F., and J. Alessi. 1971. Nitrogen fertilization of semiarid grasslands: plant growth and soil mineral N levels. Agron. J. 63:277–280.

Rauzi, F., R.L. Lang, and L.I. Painter. 1968. Effects of nitrogen fertilization on native rangeland. J. Range Manage. 21:287–291.

Seaker, E.M., and W.E. Sopper. 1988. Municipal sludge for minespoil reclamation: I. Effects on microbial populations and activity. J. Environ. Qual. 17:591–597.

Smoliak, S. 1965. Effects of manure, straw and inorganic fertilizers on northern great plains ranges. J. Range Manage. 18:11–15.

Stevenson, B.G., C.M. Parkinson, and M.J. Mitchell. 1984. Effect of sewage sludge on decomposition processes in soils. Pedobiologia 26:95–105.

U.S. Department of Agriculture, Soil Conservation Service. 1988. Soil survey of Socorro County Area, New Mexico. U.S. Gov. Print. Office. Washington, DC.

Whitford, W.G., E.A. Aldon, D.W. Freckman, Y. Steinberger, and L.W. Parker. 1989. Effects of organic amendments on soil biota on a degraded rangeland. J. Range Manage. 42:56–60.

Wight, J.R., and A.L. Black. 1979. Range fertilization: Plant response and water use. J. Range Manage. 32:345–349.

31 A Nitrogen Budget Prediction of Nitrate in Ground Water at a Farm Using Sewage Sludge

M. J. Goss
D. A. J. Barry
D. Goorahoo
P. S. Smith

Centre for Land and Water Stewardship
University of Guelph
Guelph, Ontario, Canada

There is interest in developing a closed loop approach to solving problems associated with the management of organic wastes. In this context, agriculture produces materials that are primarily consumed in urban areas. A by-product of the consumption is sewage sludge, which increasingly presents a problem for waste management. As crop production is considered to reduce the nutrient and organic matter content of soil, a solution often advocated is to spread sewage sludge on agricultural land. This helps to replace materials removed in plant produce, and simultaneously, reduces pressure on landfill sites in which sludge is commonly disposed.

If sludge is used as a source of nutrients, one problem is associated with the development of guidelines for appropriate application rates. The metal content of sludge is often a major consideration limiting the cumulative amount of material that can be applied to a field. Plant nutrients are commonly present in both mineral and organic forms in sludge. Hence, application rates need to be selected so that sufficient nutrients are provided to the crop without leaving residues that can result in contamination of water resources. This is further complicated because the ratio by weight of nutrients in sludge is commonly quite dissimilar to the requirements of crop plants. Crops commonly take up nitrogen (N), phosphorus (P) and potassium (K) approximately in the ratio 10:1:10 (Barraclough, 1989), but in sludge the ratio may be closer to 10:10:1 (Soon et al., 1978; Pierzynski, 1994). If application rates are based on the N requirements of the crop, an excess of P is likely. A more satisfactory solution is the augmentation of sludge with other fertilizers to provide an appropriate balance of nutrients. This can be achieved more readily if sludge is applied at particular times during the crop rotation.

Much early effort was aimed at optimizing application rates to meet crop yield goals. The current need, however, is to evaluate management practices within the context of the whole farming system, and assess any impacts on the

Sewage Sludge: Land Utilization and the Environment. SSSA Miscellaneous Publication.

environment. One way to study the effectiveness of current practices is to compare predictions of their impact on water quality with measured values. Limitations in our understanding of soil and plant processes can then be identified, and possibilities for extending recommendations on fertilizer practices to other soils and climates can be increased.

This chapter illustrates the use of a simplified N budget for estimating the contamination of water resources with nitrate (NO_3) and illustrates its use for assisting in developing advice on the use of sewage sludge as a nutrient source.

METHODOLOGY

Simplified N budgets can be constructed for any farm. One requirement is knowledge of purchases of N–containing materials and produce, and reliable estimates or measurements of natural inputs. These inputs are summarized in Fig. 31-1.

In Ontario, as in many other regions, a major input of N comes from the fixation of atmospheric N_2 by grain legumes such as soybeans [*Glycine max* (L.) Merr.]. This input component can be estimated from crop yield (Fig. 31-2). Nitrogen in livestock wastes can be estimated from the weight of animals bought and sold.

When the farming practices have been employed for many years, it can be assumed that processes governed by a farming system will approach an equilibrium. For example, the soil organic matter will have reached a steady state value, which may change slightly from season to season, but which will not vary greatly from one cycle of the crop rotation to another. In this case the inputs of N to the farm will be balanced by outputs, including losses to the environment. The outputs are summarized in Fig. 31-3.

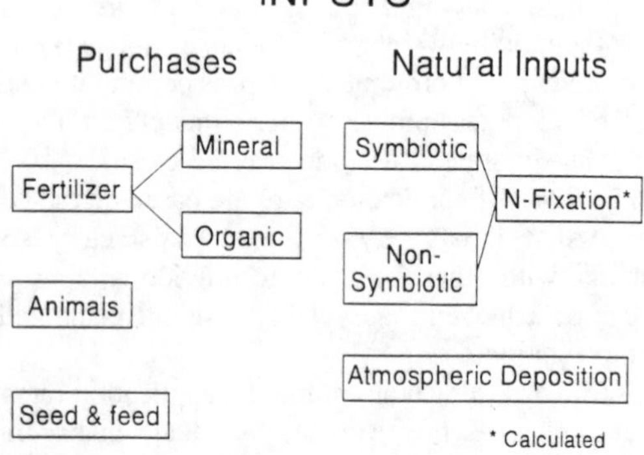

Fig. 31-1. Nitrogen inputs to farms.

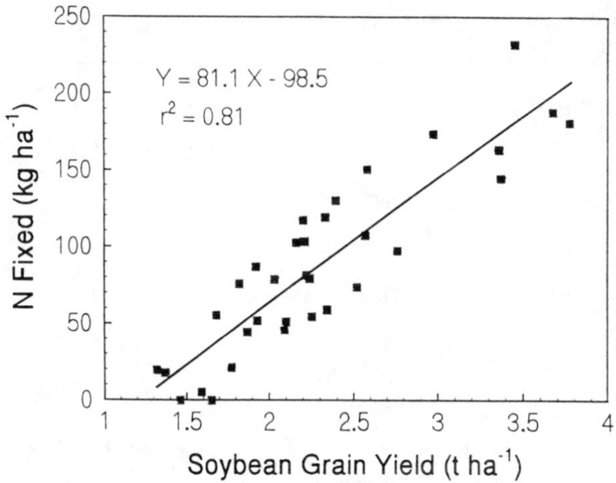

Fig. 31-2. Relationship for estimating symbiotic nitrogen fixation from grain yield of soybeans. Results from field experiments in Ontario.

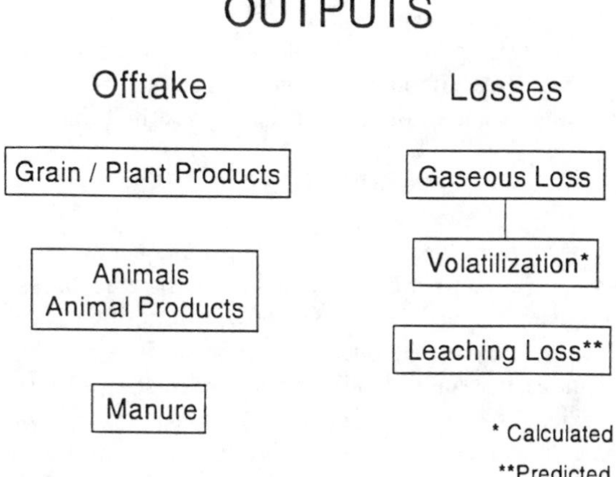

Fig. 31-3. Nitrogen outputs from farms.

The excess of inputs of N over estimated outputs was assumed to be available for leaching. The average annual recharge of ground water in Ontario has been estimated to be equivalent to 160 mm rainfall. The concentration of N in the ground water was then predicted assuming that the excess N was present as NO_3 and was dissolved in the ground water recharge. A full account of the method used in developing simplified N budgets, and the subsequent prediction of NO_3 leaching is given in Barry et al. (1993).

Table 31-1. Details of the farm where sewage sludge was applied, and a summary of the production system.

Size of farm (ha)	40.5
Soil type	Sandy loam
Rotation	Tomato†/Seed corn‡
Fertilizer N rates (kg N ha^{-1})	
Seed corn	202
Tomato	45
Sludge after tomato (L ha^{-1})	34000
Average yields (Mg ha^{-1})	
Seed corn	11.3
Tomato	47.7
Depth of water well (m)	9.0
Date of well construction	1979

†*Lycopersicon lycopersicum* (L.) Karsten.
‡*Zea mays* L.

In Ontario we have surveyed the water quality of ≈1300 farm wells. Sewage sludge had been applied on 26 of these farms, and we had sufficient information to construct a simplified N budget for one of these farms. The details of the farming practices on this farm are given in Table 31-1.

RESULTS

The survey of farm wells showed that the frequency of contamination with NO_3 and bacteria was similar to that on comparable farms where only mineral fertilizers were applied (Table 31-2). The simplified N budget for the farm showed N inputs exceeded outputs by almost 50 kg N ha^{-1} yr^{-1}, even though sewage sludge was only applied on a 5-yr rotation (Table 31-3). The excess N was predicted to result in ground water contamination with NO_3 up to 29 mg N L^{-1}. The concentration actually measured in the farm well was greater than the 10 mg N L^{-1} acceptable for drinking water, but was less than the predicted value. There was, however, a significant increase in the concentration between the two sampling occasions.

We recalculated the budget making the assumption that no sewage sludge had been applied, and that the fertilizer application was the same as in those years when sludge was not applied (Table 31-4). This resulted in the prediction that leaching would be less if sludge were not applied. The concentration of NO_3 in the well water, however, would still have exceeded the maximum acceptable concentration for drinking water.

Table 31-2. Effect of sludge application on contamination of farm wells with nitrate and bacteria.

Fertilizer	Wells	Exceed MAC†	
		NO_3	Bacteria
	no.	%	%
Mineral fertilizer only	107	19.6	29.0
Sludge applied	17	11.8‡	29.4‡

†Maximum acceptable concentration.
‡Not significant.

Table 31-3. Whole farm N budget for a tomato–seed corn rotation where sewage sludge is applied after tomato once every 6 yr.

N Inputs		N Outputs	
Source	Amount of N	Source	Amount of N
	$kg\ ha^{-1}\ yr^{-1}$		$kg\ ha^{-1}\ yr^{-1}$
Seed	1.0	Plant produce	116.6
Feed	0	Manure	0
Fertilizer	112.4	Animals	0
Manure (sludge)	28.1	Animal produce	0
Animals	0	Volatilization†	1.4
N_2 fixation‡	5.0		
Atmosphere	18.4		
Total	164.9	Total	118.0
Inputs minus outputs			$46.9\ kg\ ha^{-1}\ yr^{-1}$
Ground water recharge			$160.0\ mm\ yr^{-1}$
Predicted N in ground water			$29.3\ mg\ L^{-1}$
Measured NO_3 in well water (Winter 1991)			$13.3\ mg\ N\ L^{-1}$
(Summer 1992)			$16.9\ mg\ N\ L^{-1}$

†5% of sewage sludge N.
‡Symbiotic = 0 and nonsymbiotic = 5.0 $kg\ N\ ha^{-1}\ yr^{-1}$.

Table 31-4. Whole farm N budget for a tomato–seed corn rotation assuming no sewage sludge applied.

N Inputs		N Outputs	
Source	Amount of N	Source	Amount of N
	kg ha^{-1} yr^{-1}		kg ha^{-1} yr^{-1}
Seed	1.0	Plant produce	116.6
Feed	0	Manure	0
Fertilizer	123.7	Animals	0
Manure (sludge)	0	Animal produce	0
Animals	0	Volatilization	0
N_2 fixation†	5.0		
Atmosphere	18.4		
Total	148.1	Total	116.6
Inputs minus outputs		31.5 kg ha^{-1} yr^{-1}	
Ground water recharge		160.0 mm yr^{-1}	
Predicted N in ground water		19.7 mg L^{-1}	

† Symbiotic = 0 and nonsymbiotic = 5.0 kg N ha^{-1} yr^{-1}.

CONCLUSIONS

The simplified N budget gave a useful indication that there was excess N entering the farming system. It indicated that sludge application was increasing the potential for NO_3 leaching.

REFERENCES

Barraclough, P.B. 1989. Root growth and nutrient uptake by field crops under temperate conditions. Ann. Appl. Biol. 22:227–233.

Barry, D.A.J., D. Goorahoo, and M.J. Goss. 1993. Estimation of nitrate concentrations in ground water using a whole farm nitrogen budget. J. Environ. Qual. 22:767–775.

Pierzynski, G.M. 1994. Plant nutrient aspects of sewage sludge. p. 21–25. *In* C.E.Clapp et al. (ed.) Sewage sludge: Land utilization and the environmenmt. SSSA Misc. Publ. SSSA, Madison, WI (this publication).

Soon, Y.K., T.E. Bates, and J.R. Moyer. 1978. Land application of chemically treated sewage sludge: I. Effects on crop yield and nitrogen availability. J. Environ. Qual. 7:264–269.

32 Effect of Twelve Years of Liquid Digested Sludge Application on the Soil Phosphorus Level

A. E. Peterson
P. E. Speth
R. B. Corey
T. H. Wright

Department of Soil Science
University of Wisconsin-Madison
Madison, Wisconsin

P. L. Schlecht

Milwaukee Metropolitan
Sewerage District
Oak Creek, Wisconsin

An old Chinese proverb states "Man belongs to the soil, not the soil to man." For centuries, the Chinese have realized the value of returning organic waste to the land, and today >50% of the fertility required for growing their crops comes from such sources. In his book entitled *Farmers of Forty Centuries*, F. H. King (1911), studied the land use of organic waste in the Far East early in the 20th century and emphasized the value of such material for supplying the necessary elements for plant growth.

The disposal of sewage sludge is a situation that man has been faced with for quite some time, but the problem is becoming greater as population increases and laws governing the methods of disposal become more stringent and costs increase. Energy conservation begins with recycling. Sewage sludge application to cropland is one such opportunity. The 1977 Federal Clean Water Act and its subsequent amendments insure increasing amounts of sewage sludge as treatment plants improve their methods. This sewage sludge can be: (i) buried in a landfill where it will produce methane (sewer gas) for many years and may cause an explosion such as the one that occurred in Madison, WI in October 1983; (ii) incinerated using some fuel (oil or natural gas) to destroy this valuable organic matter and nitrogen (N) while creating a potential air pollution situation (the ash must still be disposed of in a landfill site); or (iii) perhaps the best option, land application according to recommendations established by the U.S. Environmental Protection Agency (USEPA) and state governments (Peterson et al., 1982). Use of sewage sludge on land also recycles valuable plant nutrients.

Sewage Sludge: Land Utilization and the Environment. SSSA Miscellaneous Publication.

The land application of sewage sludge not only reduces the farmer's fertilizer cost, but in most cases, reduces the disposal cost to the municipality. Most agronomists are convinced that a well-managed sludge utilization program, application of these sludges to cropland, has the lowest environmental hazard of any of the sludge disposal options (Wei et al., 1985; Peterson et al., 1988). Nationwide, somewhere between 1 to 2% of the cropland would be needed if all sewage sludge were land applied. Some may still be concerned, however, about environmental factors associated with land application of some sludge such as the effect of the large annual phosphorus (P) applications on the uptake of plant nutrients and on the environment.

HISTORICAL

More than 70 yr ago, the Milwaukee Metropolitan Sewerage District in 1923 published a bulletin on the land application of such organic wastes. Thus was born the material known as Milorganite®. Work at the University of Wisconsin for the next 30 yr explored much of the basic chemistry necessary to understand the release of elements from Milorganite® and their uptake by plants. Today such basic information is even more necessary. The Department of Soil Science at the University of Wisconsin-Madison, has studied the effect of sludge on crop production for >50 yr. Noer and Truog (1926) studied the value of activated sludge as a fertilizer. Muckenhirn (1936) evaluated the boron (B), copper (Cu), and manganese (Mn) supplying power of Milorganite®, while Rehling and Truog (1940) studied the use of Milorganite® as a source of minor elements. In recent years, Kelling et al. (1977) studied the effect of field applications of Zanesville, Wisconsin, sludge on Plano (fine-silty, mixed, mesic Typic Argiudoll) silt loam and Warsaw (fine-loamy over sandy or sandy-skeletal, mixed, mesic Typic Argiudoll) sandy loam soils. Walsh et al. (1976) summarized these findings and provided background information for sludge application guidelines (Keeney et al., 1975).

State and federal laws require that communities treat wastewater through a two- or three-step process involving primary, secondary, or tertiary treatments. Phosphorus removal techniques require adding ferric chloride or aluminum sulfate to precipitate the P as ferric or aluminum phosphates; thus increasing amounts of P will be found in the sludge. The end products of all sewage treatment processes are sewage effluent and sewage sludge. The sewage effluent is essentially clear water that contains all the soluble materials, including low concentrations of plant nutrients and other inorganic matter and traces of organic matter. This is chlorinated and discharged into a river or stream and is ready for reuse. More than 99% of the incoming wastewater will leave the treatment plant in this manner. Sewage sludge contains the solids remaining after treatment and contains substantial amounts of plant nutrients and other insoluble elements. Usually sludge contains ≈3 to 5% solids on a dry-weight basis, but if dewatered, may contain 20% or more solids. A quantity of 103 000 L (1 acre-in) of 3% solids contains ≈225 kg of N, 68 kg of P, and 23 kg of potassium (K), heavy metals, and insoluble organics.

The USEPA has been required by Congress to update their regulations on handling sewage sludge. After many delays, the National Sewage Sludge Rule (40 CFR Part 503) was signed by USEPA Administrator William K. Reilly on 25 Nov. 1992 and appeared in the 19 Feb. 1993 Federal Register. This rule had been 15 yr in the making, and at a press conference on 1 Dec. 1992 USEPA Deputy Assistant Administrator for Water, Martha Prothro stated:

> "Now let me turn to the rule itself and talk about what EPA has learned after 15 years of research and study. The result is actually good news. Sewage sludge is a product of removing solids from the wastewater discharged to municipal sewage treatment facilities. Almost all sewage sludge can be used safely on our farms, gardens, golf courses, lawns, and forests. Municipal sludge has actually been used in these ways in the U.S. for many years. Sludge has even been used to reclaim abandoned strip mines and at least one Superfund site. Of course, contaminated, untreated and improperly handled, sludge can pose health and environmental risks. That is why this rule is necessary even for valuable beneficial uses of sludge: to ensure first that the overall quality of sewage sludge used on our land remains safe; and secondly, that it is placed on the land in a safe manner."

Thus, land application is likely to increase, and it is up to us to see that sludge is applied in a safe manner. Modern techniques produce sewage sludge that often contains about the same amount of available N and total P. Since annual sludge applications are based on the amount of available N that the next crop will require, extra P will necessarily be applied. There is no question that large amounts of P are accumulating at normal sludge applications rates. The question is, is that bad, and if so, how bad? What effect does it have on crops and on the environment? In the late 1970s, a study on P from various sludge and chemical sources at different application rates was conducted (Corey, 1992), which looked at effects of applications on crop yields and P content in terms of the [aluminum (Al) + iron (Fe)] to P ratio because it is the Fe and Al in sludge treatment that usually precipitates the P. The more Fe and Al, the more you precipitate. Actually you are forming a mixed mess, a solid solution rather than a specific crystalline PO_4. The more Fe in relation to PO_4 in that solid solution, the less soluble the PO_4 in the system. To compare sludge P to chemically precipitated forms, $AlPO_4$, $FePO_4$, as well as monocalcium phosphate [$Ca(H_2PO_4)_2$, (the type of PO_4 found in most fertilizers)] was added. The sludges used varied widely in P content and (Al + Fe) to P ratio (Table 32-1).

In a field study, Corey (1992) applied 3800 kg ha^{-1} of P in Oshkosh Zimpro sludge on a red clay soil in Winnebago County (Table 32-2). This particular soil had a reasonably good P content to begin with. The effect of the 22.3 Mg ha^{-1} of sludge (3800 kg P ha^{-1}) on the P content of corn (*Zea mays* L.) leaf tissue was minimal. The sorghum-sudan [(*Sorghum bicalor* (L.) Moench] showed just a small increase. The P concentration in the corn ear leaf and grain remained the same. This was the Oshkosh sludge that in Table 32-1 showed about the third highest P availability of all the sludges that were used. It did

Table 32-1. Effect of (Al + Fe) to P ratio in sludge on P concentration in soybean [*Glycine max* (L.) Merr.] tissue grown on soil–sludge mixtures containing the equivalent of 6048 kg ha^{-1}(Corey, 1992).

Sludge or P source	(Al + Fe) to P	P in tissue
		%
$Ca(H_2PO_4)_2$	--	1.62
$AlPO_4$	--	0.41
$FePO_4$	--	0.33
Madison	0.6	0.69
Sheboygan Falls	1.4	0.42
Two Rivers	1.6	0.39
Milwaukee South Shore	1.7	0.21
Oshkosh	2.1	0.51
Fond du Lac	2.3	0.32
Wisconsin Rapids	2.4	0.32
Chicago WSW Plant	3.1	0.32
Control soil	--	0.12

Table 32-2. Concentration of P in 1977 sorghum–sudan tissue, 1978 corn ear leaf tissue and grain, and 1980 corn grain from Oshkosh sludge plots, Winnebago Co., Wisconsin (Corey, 1992).

	1977		1978		1980
Sludge rate	P in sorghum–sudan cut 1	cut 2	P in corn ear leaf	P in corn grain	P in corn grain
Mg ha^{-1}			%		
0	0.35	0.47	0.25	0.32	0.37
2.2	0.36	0.49	0.25	0.32	0.41
4.4	0.35	0.51	0.26	0.33	0.36
8.8	0.36	0.53	0.24	0.32	0.37
22.3	0.37	0.55	0.26	0.32	0.41

not increase the availability of P significantly, and therefore, should not contribute to a nutrient imbalance. On the other hand, the P will not leach out so this application is going to maintain that concentration of P almost forever. Even by cropping it down, one would probably never run out of P for many decades (Peterson & Corey, 1993).

This all means that P in sludge should not hurt the balance of nutrients in the plant. The one exception is zinc (Zn), and we know that if we increase P to a high level, it would tend to depress the uptake of Zn and might result in Zn deficiency. With sludge, however, you are adding Zn at the same time that you add the P so you should never get into trouble from P-induced Zn deficiency. Thus, as long as you do not exceed the N requirements of the crop, you are not going to get any problems with crop yield or composition because of your sludge applications.

PRESENT RESEARCH

The research proposal *Utilization of Wastewater Solids from the Metropolitan Sewerage District on Cropland* was submitted to the Milwaukee Metropolitan Sewerage District by the Department of Soil Science in March 1979, and approved by the District in June 1979. The desire to perform ground water studies necessitated the use of relatively large plots, which in turn, required the use of field-scale equipment. Researchers selected the Lakeland Farm as the test site since the Plano silt loam (prairie soil) at this site is characteristic of this upland soil in southeastern Wisconsin. The initial soil test indicated: cation-exchange capacity, 17 $cmol_c$ kg^{-1}; pH, 6.6; organic matter, 21 Mg ha^{-1}; P, 98 mg kg^{-1}; K, 165 mg kg^{-1}; Ca, 2450 mg kg^{-1}; magnesium (Mg), 410 mg kg^{-1}; Zn, 8 mg kg^{-1}.

The overall objective of this 12-year study is to investigate the effects of utilization of wastewater solids on agricultural land. The specific objectives are: (i) to determine the effect of initial and annual sludge applications on plant growth; (ii) to determine the plant recovery and utilization of N and P in the sludge; (iii) to determine the influence of sludge on levels of chemical contaminants in plants and determine whether any contaminants move into the subsoil; (iv) to determine the movement of nitrates or other potentially harmful elements to the ground water; and (v) to determine the influence of sludge applications on the physical condition of the soil. Initially, one objective was to determine the effects of two modes of application (direct injection vs. surface application–incorporation) on any of the above. However, no differences were noted after 4 yr, and only injection has been used since then.

PROCEDURE

Fourteen test wells were installed in the fall of 1979. Nine additional wells were installed in April 1984. Initial Milwaukee South Shore Wastewater Treatment Plant sludge applications of 0, 6.6, and 13.2 Mg ha^{-1} of dry matter were applied between 24 to 27 Sept. 1979 and annual applications of similar rates on the north half of each plot were made in November, except in 1988 when the drought limited corn growth to that of the untreated plots; thus, no additional N was applied for the 1989 growing season. All plots have received broadcast applications of 180 kg ha^{-1} of 0–0–60 (N–P–K) in November 1981, April 1984, April 1986, and April 1990; 13.2 Mg ha^{-1} of lime was applied to all plots in May 1984, and worked into the soil prior to planting the corn. In the spring of 1984, a fertilizer-only treatment (equivalent to the 6.6 Mg ha^{-1} sludge treatment was established using the plots that had received the 6.6 Mg ha^{-1} application in the fall of 1979 (one time only).

Corn is grown as the indicator crop and is removed for silage. Sludge from the Milwaukee South Shore Wastewater Treatment plant is injected in late October or November. Except for fertilizer, the plots receive the same treatment as other corn fields on the Lakeland County Farm. The variety is usually a 105- to 110-d relative maturity, and the recommended weed control practices are used. According to Mr. Al Wood, farm manager, in 1992 this consisted of a 110-d relative maturity corn planted in 30-inch rows at 79,000 kernels ha^{-1}, and a preplant herbicide application of about 5 L ha^{-1} Lasso. Because of the dry weather in 1992, however, the herbicides were not very effective, and in spite of cultivation, some areas had reduced growth because of weed competition.

RESULTS

Crop Yields

Corn grain yields (Table 32-3) indicate the fertility value of sewage sludge providing there is adequate moisture, with 1988 the exception. The nutrients carried over to 1989, since yields again increased without any sludge or fertilizer application.

Soil Analysis

Soil samples to a depth of 92 mm are taken in May, July, and October. The soil pH has risen ≈0.3 of a unit (6.5 to 6.8), since the lime application of 1984. The normal sludge rate (6.6 Mg ha^{-1}) supplied the corn with ≈196 kg ha^{-1} of available N and P. After 12 yr of corn production and annual sludge applications of 6.6 and 13.2 Mg ha^{-1}, the soil P level tested 175 and 180 mg kg^{-1}, respectively (Table 32-4). The results have shown that the high soil P is not hurting the balance of nutrients in the plants. Thus, as long as you do not exceed the N requirements of the crop, there should not be any problems with crop yield or composition because of the P in the sludge application.

Plant Analysis

Corn has been grown as the indicator crop. The ear leaf (sampled at tasseling) and the grain P content has only varied between 0.3 to 0.4% on *all* samples for all years in spite of the high soil P levels. This is to be expected since Corey (1992) has shown that application of 3800 kg $ha^{-1}$of P from sewage sludge had minimal effect on the P content of corn ear leaf and grain. The Zn content of the ear leaf and grain is given in Table 32-5.

Table 32-3. Milwaukee Metropolitan Sewerage District sludge experiment: Corn grain yields[†] in Elkhorn, WI.

Sludge application rate[‡]	Yield													
	1980	1981	1982	1983	1984	1985	1986	1987	1988	1989	1990	1991	1992	1993
$Mg\ ha^{-1}$	kg ha^{-1}													
0	9971	7838	7713	7776	5080	6898	4766	7024	4766	5769	5644	3888	6020	4578
6.6 (one-time)	9469	8779	8654	8090	4202	7462	6083	7212	4891	5957	6208	4390	6083	4879
13.2 (one-time)	9971	8717	8717	8466	4013	8152	6334	7086	4703	6772	6208	4577	6647	5142
6.6 (annual)	--	10 284	9281	10 096	8027	9845	8968	9468	4829	8340	10 221	7274	10 473	10 159
13.2 (annual)	--	10 912	9093	9971	7839	9907	9093	9218	4264	8591	10 598	7024	10 034	10 096
Fertilizer	--	--	--	--	6961	9030	8905	9783	4954	8340	10 159	7462	10 413	9908

[†]Average of three replicates; 15% moisture.
[‡]On a dry-solids basis.

Table 32-4. Milwaukee Metropolitan Sewerage District sludge experiment: Soil P analysis in Elkhorn, WI.

Treatment	Soil P†										
	1981	1982	1983	1984	1985	1986	1987	1988	1989	1990	1991
$Mg\ ha^{-1}$	$mg\ kg^{-1}$										
0	89	68	53	48	60	54	58	63	50	45	78
6.6	98	74	89	130	130	112	135	145	150	150	175
13.2	150	125	110	100	195	180	195	190	195	195	180
Fertilizer	--	--	--	78	99	100	130	145	120	135	200

† Initial soil analysis (1979) = 98 $mg\ kg^{-1}$.

Table 32-5. Milwaukee Metropolitan Sewerage District sludge experiment: Zn analysis of corn tissue in Elkhorn, WI.

	Zinc concentration										
Treatment	1981	1982	1983	1984	1985	1986	1987	1988	1989	1990	1991
$Mg\ ha^{-1}$						$mg\ kg^{-1}$					
						Ear leaf					
0	26	25	29	24	24	21	34	34	38	18	24
6.6	48	33	60	71	57	56	95	105	90	55	82
13.2	68	42	70	92	88	82	114	116	107	73	122
Fertilizer[†]	--	--	--	37	40	27	37	37	47	23	33
						Grain					
0	15	17	19	21	16	18	17	16	13	20	21
6.6	17	16	23	24	17	24	22	23	18	22	24
13.2	22	17	24	24	24	18	26	21	23	25	36
Fertilizer[†]	--	--	--	22	15	16	16	18	12	12	17

[†]Fertilizer plots started in 1984.

Plant analyses have indicated that all the major plant nutrients fall within the sufficient range and are above the critical level regardless of sludge application. The increased Zn content of the ear leaf tissue (Table 32-5) from all sludge applications indicates a possible benefit from the Zn contained in the sludge. Zinc has been recognized as the minor element most likely to be limiting in growth of field corn on high organic matter soils.

CONCLUSIONS

There seems to be no reason to limit sludge-P application because of any effect on the crop. Also, because the solubility of the phosphate in sludge is not very high, we have a low soil solution concentration. We do not need to worry about P leaching to ground water because leaching is practically zero. The only thing left that is potentially detrimental is erosion of the high-P sediment. Significant eutrophication in surface waters may occur if you erode high-P soil particles off your land into the surface waters. If we were to limit the P application of sludge because of the erosion hazard, however, we would have the ripple effect in other areas. First of all, this would limit the P application to ≈33 kg ha^{-1} if the growing plant is to remove as much P as is being applied so that the soil test will be maintained below 336 kg ha^{-1}. Thus, only 33 kg ha^{-1} instead of 270 kg ha^{-1} P or only one-eighth of the present application that is now allowed. Therefore, land application of the sludge would require eight times more acreage. Finding more land means greater hauling distances. This means greater costs because you are applying less in one place, driving to more farms, requiring more permits and more road upkeep. Also, land that is suitable for sludge applications may run out, which would force municipalities to go to some alternative method of sludge disposal.

The comparative risk assessments for methods of sludge disposal show land spreading to have the least risk. Incineration and landfilling have a much higher risk. Therefore, P limits could force treatment plants into a disposal method with greater risks.

Greater risks would have to be weighed against the benefits derived from not landspreading sludge. It seems much more practical to put additional effort into erosion control practices such as increased setbacks from waterways and good conservation tillage practices and keep the land requirements essentially as they would be under the new (relaxed) 503 regulations. Limiting P applications to stay below a 150 mg kg^{-1} soil test level could force many treatment plants, particularly the larger ones, into less-desirable, higher-risk methods of disposal.

It is important to realize that everyone pays a part of the cost of disposal of our waste, whether it is solid or liquid. Landfills will fill much more rapidly if digested sewage sludge is placed in these locations. The farmer who is wise enough to have sludge applied on his cropland will not only gain the benefits of the nutrients being added, but will also be saving money for the taxpayers. It is time we realize that sewage sludge will not evaporate and that society as a whole will benefit from wise use of this material, and thus, recycle a valuable resource. Let's not go into the *zero risk syndrome*, and forget the basic facts about living in a more complicated environment.

REFERENCES

Corey, R.B. 1992. Phosphorus regulations: Impact of sludge regulations. Crops Soils Newsl. 20:5–10.

Keeney, D.R., K.W. Lee, and L.M. Walsh. 1975. Guidelines for the application of wastewater sludge to agricultural land in Wisconsin. Wisconsin Depart. of Natural Resources Tech. Bull. 88, Madison.

Kelling, K.A., A.E. Peterson, L.M. Walsh, J.A. Ryan, and D.R. Keeney. 1977. A field study of agricultural use of sewage sludge - effect on crop yield and uptake of N and P. J. Environ. Qual. 6:339–345.

King, F.H. 1911. Farmers of forty centuries. Rodale Press, Emmaus, PA.

Muckenhirn, R.J. 1936. Response of plants to boron, copper and manganese contained in "Milorganite." J. Am. Soc. Agron. 28: 829–842.

Noer, O.J., and E. Truog. 1926. Activated sludge: Its production, composition, and value as a fertilizer. J. Am. Soc. Agron. 18: 953–962.

Peterson, A.E., and R.B. Corey. 1993. Implications of soil phosphorus buildup from repeated sewage sludge applications. p. 88–97. *In* Proc. 1993 Fert., Aglime, and Pest Mgmt. Conf. Madison, WI. Jan. 1993. Univ. of Wisconsin Extension Service, Madison.

Peterson, A.E., R.B. Corey, K.A. Kelling, and D. Taylor. 1982. Sewage sludge application to Wisconsin cropland--an update. p. 60–70. *In* Proc. 1982 Fert., Aglime, and Pest Mgmt. Conf. Madison, WI. 18-20 Jan. 1982. Univ. of Wisconsin Extension Service, Madison.

Peterson, A.E., P.E. Speth, and P.L. Schlecht. 1988. Effect of sewage sludge applications on groundwater quality. p. 272–285. *In* Proc. 11th Ann. Madison Waste Conf., Madison, WI. 13-14 Sept. 1988. Univ. of Wisconsin, Madison.

Rehling, C.J., and E. Truog. 1940. "Milorganite" as a source of minor nutrient elements for plants. J. Am. Soc. Agron. 32:894–906.

Walsh, L.M., A.E. Peterson, and D.R. Keeney. 1976. Sewage sludge--wastes that can be resources. Wisconsin Agric. Res. Stn. Rep. R2779. Madison.

Wei, Q.F., B. Lowery, and A.E. Peterson. 1985. Effect of sludge application on physical properties of a silty clay loam soil. J. Environ. Qual. 14: 178–180.

Participant List

Larry Adams
USDA
324A Administration Bldg.
14th & Independence Ave. SW
Washington, DC 20250-0111
202-205-5853

Richard Aguilar
USDA-Forest Service
Rocky Mountain Station
2205 Columbia SE
Albuquerque, NM 87106
505-766-1045

J. Scott Angle
Dep. of Agronomy
University of Maryland
College Park, MD 20742
301-405-1346

Dean Barry
University of Guelph
Richards Bldg.
Guelph, Ontario N1G 2W1
Canada
519-824-4120, Ext. 4263

Nick Basta
Dep. of Agronomy
Oklahoma State University
Stillwater, OK 74078
405-744-9568

Robert K. Bastian
USEPA
OWEL (WH-547)
401 M St. SW
Washington, DC 20460
202-260-7378

Karin Beaton
University of Adelaide
Waite Road
Glen Osmond, Adelaide
5064 South Australia
Australia 08/3037285

Lisa Beggs
Allegheny College
Box 447
Meadville, PA 16335
814-445-9757

Arnold Blachanski
City of Sauk Centre
Rt. 2 Box 125
Sauk Centre, MN 56378
612-352-6723

Herschel Blasing
MPCA
1601 Minnesota Dr.
Brainerd, MN 56401
218-828-6064

Maura Bracken
MPCA
520 Lafayette Rd.
St. Paul, MN 55104
612-296-7376

Gary C. Brandt
City of Lincoln
2400 Theresa St.
Lincoln, NE 68521
402-441-7968

Robert Bricker
Dep. of Plant and Soil Science
1066 Agric. Sci. Bldg.
West Virginia University
Morgantown, WV 26506
304-293-2219

Pat Burford
Metropolitan Waste Control Commission
2400 Childs Rd.
St. Paul, MN 55106-6724
612-772-7379

Stephen Burke
Allegheny College
Box 529
Meadville, PA 16335
603-776-6917

Bonnie Jean Cameron
Ag-Chem Equipment Co.
5720 Smetana Dr., Ste. 100
Minnetonka, MN 55343-9688
612-933-9006

Rufus L. Chaney
USDA-ARS
Environmental Chemistry Lab
Bldg. 007 BARC-West
Beltsville, MD 20705
301-504-8324

H. H. Cheng
Dep. of Soil Science
University of Minnesota
St. Paul, MN 55108
612-625-9734

Rob Christy
RDP Co.
531 Plymouth Rd.
Plymouth Meeting, PA 19462
215-941-9080

C. Ed Clapp
USDA-ARS
Dep. of Soil Science
University of Minnesota
St. Paul, MN 55108
612-625-2767

Phil Cochran
Cochran Agronomics
Rt. 6 Box 200
Paris, IL 61944
217-465-5282

Donald Crossett
Ministry of the Environment and Energy
RR 7
Tillsonburg, Ontario
Canada
519-485-2860

Margaret Crossett
Kon-Mag
Ayr, Ontario
Canada
519-740-6551

Gerry Croteau
E & A Environmental Consultants
18912 N. Creek Pkwy.
Bothell, WA 98011
206-485-3219

W. Lee Daniels
Dep. of Crop & Soil Environ. Sciences
244 Smyth Hall
Virginia Tech
Blacksburg, VA 24061-0404
703-231-7175

Paul E. Dohlman
Iowa Rural Water Association
100 Court Ave., Ste. 205
Des Moines, IA 50309
515-283-8214

Bob Dowdy
USDA-ARS
Dep. of Soil Science
University of Minnesota
St. Paul, MN 55108
612-625-7058

Margaret DuBois
Dep. of Soil Science
University of Minnesota
St. Paul, MN 55108

Steven G. Duerre
MPCA
520 Lafayette Rd.
St. Paul, MN 55155
612-296-9264

Jorja DuFresne
MPCA
520 Lafayette Rd.
St. Paul, MN 55155
612-296-9292

El-Sayed H. M. El-Haddad
U.S. Salinity Laboratory
4500 Glenwood
Riverside, CA 92501
909-369-4834

Gamal Hassan Abd El-Hay
College of Agriculture
Al-Azhar University
Nasr City, Cairo
Egypt 850060

Gary Elsner
University of Minnesota
1299 130th Lane NE
Blaine, MN 55434

David L. Eskew
Automated Sciences Group
Suite A-300
800 Oak Ridge Turnpike
Oak Ridge, TN 37830
615-482-6601

Shawn Esser
Marathon County
Courthouse, 500 Forest St.
Wausau, WI 54403
715-847-5213

Greg Evanylo
Dep. of Crop and
Soil Environ. Sciences
Virginia Tech
Blacksburg, VA 24061-0403
703-231-9739

Anne Fairbrother
USEPA
200 SW 35th St.
Corvallis, OR 97333
503-754-4606

Evaterina Georgieva Filcheva
12, L. Uoshut St.
1606 Sofia, Bulgaria
359-2-52-35-41

Jane B. Forste
Bio Gro Systems
180 Admiral Cochrane Dr.
Annapolis, MD 21401
410-224-0022

Francisco Gavi Reyes
Oklahoma State University
121 Brumley, Apt. #1
Stillwater, OK 74074
405-744-3755

Richard E. Goff
Midwest Labs
319 S. Payne
New Ulm, MN 56073
507-354-4834

Warren Gore
Dep. of Rhetoric
University of Minnesota
St. Paul, MN 55108

M. J. Goss
University of Guelph
Center for Land &
Water Stewardship
Richards Building, Rm. 140
Guelph, Ontario, Canada
519-824-4121, Ext. 2491

Del Haag
MPCA
1601 Minnesota Dr.
Brainerd, MN 56401
218-828-6063

Dan Halbach
Dep. of Agric. & Applied Econ.
University of Minnesota
St. Paul, MN 55108

Tom Halbach
Dep. of Soil Science
University of Minnesota
St. Paul, MN 55108
612-625-3135

Thomas Hansmeyer
Dep. of Soil Science
University of Minnesota
St. Paul, MN 55108

Fred Heasley
Sims Ag
8587 N. Clear Creek Rd.
Huntington, IN 46750
219-344-1701

Stan J. Henning
3208 Agronomy Hall
Iowa State University
Ames, IA 50011
515-294-7846

Bruce Henningsgaard
MPCA
520 Lafayette Rd.
St. Paul, MN 55122
612-296-9289

Chuck Henry
College of Forest Resources,
AR-10
University of Washington
Seattle, WA 98195
206-685-1915

Stephen J. Herbert
Dep. of Plant and Soil Sciences
University of Massachusetts
Amherst, MA 01003
413-545-2250

Matt Herring
University of Missouri-Extension
Box 71
Union, MO 63084
314-583-5141

Lisa Hillster
Allegheny College
Box 1298
Meadville, PA 16335

Mark Hoeft
J & M Waste Applicators
Box 116
Madelia, MN 56062
507-642-3311

Gary Hoette
University of Missouri-Extension
Courthouse, 211 E. 3rd St.
Montgomery City, MO 63361
314-564-3733

Harry Hoitink
Dep. of Plant Pathology
The Ohio State University
Wooster, OH 44691
216-263-3848

Bobby Holder
Northwest Experiment Station
Univ. of Minnesota-Crookston
Hwy 2 & 75
Crookston, MN 56716
218-281-8267

Jayson Holt
Dep. of Soil Science
University of Minnesota
St. Paul, MN 55108

Rich Hoormann
University of Missouri-Extension
260 Brown Rd.
St. Peters, MO 63376
314-279-3000

Star Hormann
Dep. of Soil Science
University of Minnesota
St. Paul, MN 55108

Loila Hunking
N-Viro Resources
231 S. Phillips Ave., Ste. 207
Sioux Falls, SD 57102
605-338-7657

Maneesha Jain
Dep. of Soil Science
University of Minnesota
St. Paul, MN 55108

Justin A. Jeffery
USDA-SCS
6120 Earle Brown Dr., Ste. 650
Brooklyn Center, MN 55430
612-566-2941

William J. Jewell
Dep. of Agric. & Biol. Eng.
Riley-Robb Hall
Cornell University
Ithaca, NY 14853
607-255-4533

Brad Joem
Dep. of Agronomy
Lilly Hall
Purdue University
W. Lafayette, IN 47907-1150
317-494-9767

Kathryn Kellogg Johnson
HCK
21041 S. Western Ave., Ste. 160
Torrance, CA 90501
310-328-0107

Joel Johnson
Dep. of Soil Science
University of Minnesota
St. Paul, MN 55108

LeeAnn Johnson
Metropolitan Waste Control Commission
2400 Childs Rd.
St. Paul, MN 55106-6724
612-772-7365

William Jokela
Dep. of Plant & Soil Science
Hills Bldg.
University of Vermont
Burlington, VT 05405
802-656-2630

Gerald Kidder
Dep. of Soil & Water Science
University of Florida
Box 110290
Gainesville, FL 32611-0290
904-392-1951

Mark King
Maine Dep. of Environmental Protection
State House Station 17
Augusta, ME 04333
207-287-2651

Charles M. Knapp
Technical Resources
712 5th St., Ste. G
Davis, CA 95616
916-756-2252

Kevin Langford
AgriBank
375 Jackson
St. Paul, MN 55101
612-282-8542

James Larsen
Dep. of Soil Science
University of Minnesota
St. Paul, MN 55108

William E. Larson
Dep. of Soil Science
University of Minnesota
St. Paul, MN 55108
612-624-8714

Bill Lauer
Dakota County Environmental Management
14955 Galaxie Ave.
Apple Valley, MN 55124
612-891-7546

Meg Layese
Dep. of Soil Science
University of Minnesota
St. Paul, MN 55108

John K. Leslie
BFI, Browning Ferris Industries
P O Box 1375
Minneapolis, MN 55440
612-921-8617

Dennis Linden
USDA-ARS
Dep. of Soil Science
University of Minnesota
St. Paul, MN 55108
612-625-6798

Ruilong Liu
USDA-ARS
Dep. of Soil Science
University of Minnesota
St. Paul, MN 55108

Terry Logan
Ohio State University
c/o N-Viro Energy Systems
3450 W. Central #328
Toledo, OH 43606
614-292-9043

Michael Ludvik
Carver County
600 E. 4th St., Box 3
Chaska, MN 55318
612-448-1217

Cecil Lue-Hing
Metro. Water Reclamation District
100 E. Erie St.
Chicago, IL 60611
312-751-5190

George N. Lutzic
New York City DEP
96-05 Horace Harding Expy.
Corona, NY 11368-5107
718-595-5048

Larry D. Maddux
Kansas State University
6347 NW 17th
Topeka, KS 66618
354-6236

Martha Mamo
Dep. of Soil Science
University of Minnesota
St. Paul, MN 55108
612-625-1798

EuDale Mathiason
MPCA, WQ, MS
520 Lafayette Rd.
St. Paul, MN 55155
612-296-7195

Barbara McCarthy
NRRI
5013 Miller Trunk Hwy.
Duluth, MN 55811
218-720-4322

Connie Minetor
MPCA
520 Lafayette Rd.
St. Paul, MN 55155
612-296-7765

Thomas Mishmash
City of Princeton
206 S. LaGrande Ave.
Princeton, MN 55377
612-389-2040

Robert D. Munson
Dep. of Soil Science
2147 Doswell Ave.
St. Paul, MN 55108
612-644-9716

Scott R. Nelson
Metro. Water Reclamation District
of Greater Chicago
Box 368
Canton, IL 61520
309-647-8200

Mike Nicholson
N-Viro Energy Systems
3450 W. Central #328
Toledo, OH 43606
419-535-6374

Michael R. Norland
U.S. Bureau of Mines
5629 Minnehaha Ave. S
Minneapolis, MN 55075-3410
612-725-4573

George O'Connor
Dep. of Soil & Water Science
University of Florida
Box 110510
Gainesville, FL 32611-0510
904-392-1804

Steve Oberle
National Soil Tilth Laboratory
Iowa State University
Ames, IA 50011
515-294-2421

Barbara Ogg
Cooperative Extension
Lancaster County
444 Cherrycreek Rd.
Lincoln, NE 68528
402-441-7180

Kristie Otterson
Dep. of Soil Science
University of Minnesota
St. Paul, MN 55108

Albert L. Page
Dep. of Soil & Environ. Sciences
University of California
Riverside, CA 92521-0424
909-787-3654

Antonio Palazzo
U.S. Army-CRREL
72 Lyme Rd.
Hanover, NH 03755
603-646-4374

James F. Parr, Jr.
USDA-ARS-NPS
10302 E. Nolcrest Dr.
Silver Spring, MD 20903-1327
301-504-5281

Arthur Peterson
Dep. of Soil Science
University of Wisconsin
Madison, WI 53706
608-262-2631

Charlotte Peverly
1016 Bradfield Hall
Cornell University
Ithaca, NY 14853
607-255-1739

John Peverly
1016 Bradfield Hall
Cornell University
Ithaca, NY 14853
607-255-1739

Gary Pierzynski
Dep. of Agronomy
Throckmorton Hall
Kansas State University
Manhattan, KS 66506
913-532-7209

Robert C. Polta
Metropolitan Waste Control Commission
2400 Childs Rd.
St. Paul, MN 55106-6724
612-772-7390

Larry Quinlivan
National Stone Association
1415 Elliot Place NW
Washington, DC 20007
202-342-1100

Eric F. Ritchie
McCain Foods
Box 97
Florenceville, NB E0J 1K0
Canada
506-392-5541

Rochelle Robideau
3M, Environ. Laboratory
935 Bush Ave., Bldg. Z-3E
St. Paul, MN 55106
612-778-7065

Sally Robinson
N-Viro Energy Systems
3450 W. Central #328
Toledo, OH 43606
419-535-6374

Steve Rogers
Cooperative Research Centre for Soil & Land Management
Private Bag No. 2
Glen Osmond, Adelaide
5064 South Australia
Australia 08/303 8451

Gary Rudolph
EG&G Florida
Box 21267, BOC-187
Kennedy Space Center, FL 32815
407-867-3300

Jim Ryan
USEPA
401 M St. SW
Washington, DC 20460
513-569-7653

Stephen Scott
Dakota County Environmental Management
14955 Galaxie Ave.
Apple Valley, MN 55124
612-891-7537

Keith Seward
Bytec Resource Management
1020 Third Ave.
Monroe, WI 53566
608-328-8200

Willard Sexauer
Enviroland
20351 Edgeton Ct.
Richmond, MN 56368-8368
612-597-3353

Bill Sheehan
Maine Dep. of Environ. Protection
State House Station 17
Augusta, ME 04333
207-287-2651

Rex Singer
School of Public Health
University of Minnesota
1444 Englewood Ave.
St. Paul, MN 55104
612-644-0535

John J. Sloan
Dep. of Agronomy
Oklahoma State University
Stillwater, OK 74078
405-744-6414

Elmer W. (Bill) Smith
IDHW-DEQ
224 South Arthur
Pocatello, ID 83204
208-236-6160

Kelly Smith
Dep. of Soil Science
University of Minnesota
St. Paul, MN 55108

Peter S. Smith
Dep. of Land Resource Sci.
University of Guelph
Rm. 137 Richards Bldg.
Guelph, Ontario N1G 2W1
Canada
519-824-4120, Ext. 4263

William E. Sopper
ERRI Land & Water Bldg.
Penn State University
University Park, PA 16802
814-863-0291

Steve Stark
Metropolitan Waste Control
Commission
2400 Childs Rd.
St. Paul, MN 55106
612-772-7358

Jeff Stein
N-Viro Energy Systems
3450 W. Central #328
Toledo, OH 43606
419-535-6374

Malcolm E. Sumner
Dep. of Crop & Soil Sciences
3111 Miller Plant Sci.
University of Georgia
Athens, GA 30602
706-542-0899

Ted Tatem
City of Calgary
Box 2100, Stn. M#37
Calgary, Alberta T2P 2M5
Canada
403-265-5575

Sue Thomas
Dep. of Soil Science
University of Minnesota
St. Paul, MN 55108

Steve Titko
O. M. Scotts
14111 Scottlawn Rd.
Marysville, OH 43041
513-644-7559

Brad Trosper
Conf. Salish & Kootenai Tribes
Box 278
Pablo, MT 59855
406-675-2700

Princesa VanBuren
3001 County Rd. #146
Clearwater, MN 55320
612-558-6330

David L. Veith
U.S. Bureau of Mines
5629 Minnehaha Ave. S
Minneapolis, MN 55417
612-725-4709

Gordon Voss
Metropolitan Waste Control Commission
2400 Childs Rd.
St. Paul, MN 55106-6724

John M. Walker
USEPA
401 M St. SW
Washington, DC 20460
202-260-7283

Pete Weidman
Western Lake Superior Sanitary District
2626 Courtland St.
Duluth, MN 55806
218-722-3336

Richard A. Weismiller
Dep. of Agronomy
University of Maryland
College Park, MD 20742
301-405-1306

Dwayne G. Westfall
Dep. of Agronomy
Colorado State University
Fort Collins, CO 80525
303-491-6149

Sue White
Dep. of Soil Science
University of Minnesota
St. Paul, MN 55108
612-625-1767

Richard A. Wiese
Extension/EPA Liaison
Dep. of Agronomy
University of Nebraska
Lincoln, NE 68583-0910
402-472-8493

Frank W. Wilson
Van Cleve Labs
322 S. 4th St.
Minneapolis, MN 55415
612-341-4511

Joe Zublena
North Carolina State Univ.
Box 7619
Raleigh, NC 27695
919-515-7302